Freeze-drying of Pharmaceuticals and Biopharmaceuticals

Freeze-drying of Pharmaceuticals and Biopharmaceuticals

Felix Franks
BioUpdate Foundation, London, UK

in association with Tony Auffret

RSC Publishing

ISBN-13: 978-0-85404-268-5

A catalogue record for this book is available from the British Library

Published by The Royal Society of Chemistry,
Thomas Graham House, Science Park, Milton Road,
Cambridge CB4 0WF, UK

Registered Charity Number 207890

For further information see our web site at www.rsc.org

Typeset by Macmillan India Ltd, Bangalore, India
Printed by Henry Lings Ltd, Dorchester, Dorset, UK

Preface

It was in the middle of the 1980s when I left the Plant Sciences Department of the University of Cambridge, to set up a start-up company on the Cambridge Science Park. For the past several years my colleagues and I had been studying the mechanisms by which plants, insects and microorganisms were able to husband their water reserves, when faced by extreme conditions of drought, freezing or salinity. We believed that we had penetrated their secrets of induced dormancy and were about to attempt to copy the physics and chemistry of ecological water stress tolerance by applying them to the *in vitro* protection of labile molecules and more complex structures; in present-day parlance: the commercialisation of natural protection against drying. I had learned much about plants by that time, but I had been the only one in Department whose activities involved the *in vivo* preservation of human red blood cells and the like.

Our venture was helped on its way with seed finance by Pafra Ltd. of Basildon, Essex, a manufacturing company, and by Trinity College, Cambridge, with the offer of suitable accommodation on their Science Park. Naively, we believed that industry would jump for joy at the advent of technology that could counter the ravages caused by water, and also by its withdrawal. We were mistaken: it took many years before the potential of our developments began to sink in. In the meantime, the Chairman of Pafra Ltd. began to ask what we might possibly do to improve our monthly negative cash flow. It was a salutary experience to one, who had worked in large educational institutions and multinational companies, where one did not have to think about where salaries came from. We put our heads together and examined our common store of knowledge that might quickly be turned into hard cash.

Having spent all my professional working life up to that time on studies that involved different aspects of Water, and having become known in the scientific community as "Water Franks", it seemed a natural starting point to think of something that involved water, perhaps its removal by currently accepted processes. This was how we hit on freeze-drying. It soon transpired that here was a technology that was universally practised by large sectors of industry, especially food and pharmaceutical processing. Of these, the latter seemed to us to be more valuable in cash terms. So we looked at current practice of what seemed a complex and daunting combination of several different components: physics, chemistry, engineering, economics, law and regulation. What struck me, personally, as strange was that there did not appear to be a single University Engineering Department in the world where freeze-drying was taken seriously. Indeed, my study of chemical engineering texts revealed that lyophilisation was never mentioned. Why not? It was not until a decade later that Biochemical Engineering became respectable.

It soon became apparent that, apart from very few places, freeze-drying, as then practised, was a push-button affair, with very little background understanding on the part of those who pushed the buttons, and even less by their managers. Life was not made any easier by the writings of the "experts" who spread the idea that "freeze-drying can only be performed on a trial-and-error basis". In the pharmaceutical industry such errors could be very costly indeed. We set about to try and remedy the situation. By applying what we knew about aqueous systems, freezing, thermodynamics and kinetics, we arrived at a synthesis of what we thought was a rudimentary understanding of the whole process from beginning to end. In retrospect it was probably very cheeky for us novices to take up the fight against the long-established "experts" and to claim that it was possible to calculate effective freeze-drying cycles without ever having to go close to a freeze-drier. We advertised our consultancy services and developed seminars throughout Europe to spread the new insights. Gradually we convinced a group of scientific colleagues of the rational basis of our approach to this well-established technology. At the end of the century we were able to count 18 of the largest 20 international pharma companies among our many clients. This book is based on the approaches developed and applied by Pafra Biopreservation, rather than trial-and-error methods, to achieve "Stability by Design".

The book is thus a distillate of our acquired knowledge and experience. Of necessity, this experience is limited to what my colleague Tony Auffret came to call the "hospital cases" of freeze-drying. Those without problems do not require the services of consultants. Our approach to the subject rests on two fundamental principles: (1) everything that a skilled engineer might contribute to the development of an efficient and economical process depends on the formulation details of the solution to be dried; in other words, every formulation is accompanied by an optimum process cycle which can be approached but not exceeded; and (2) with decreasing water content, most of the remaining products that are being dried, take on the attributes of amorphous materials and end up as glasses and must be treated as such. The realisation of basic differences between crystalline solids and glasses forms an important aspect of pharmaceutical freeze-drying.

It is ironic that, all the time we were propagating the new thinking on freeze-drying, our thoughts and endeavours were really directed towards the development of alternative pharmaceutical drying processes, superior to freeze-drying. The reasoning was that any engineer, faced with the need for removing a kilogram of liquid water, would be most unlikely to freeze the water to ice, cool the ice down to well below its melting temperature and then sublime the ice by heating it and recover the resulting water vapour by freezing it again. He would of course remove the water by boiling it.

It now requires a new educational exercise which, fortunately, has already begun, to convince the pharmaceutical industry of the merits of evaporation, rather than freezing. When I first visited the famous Materials Science Department in Cambridge, to learn about "real" glasses and spoke about water-soluble glasses, the response was one of mild disbelief. It was only by reminding the audience of sugar candy and candy floss, that suspicion turned to realisation. Now it is beginning to be understood that sugar candy and window glass are governed by the same physical laws. There can now be hope of further advances.

The first practical success of the "new" thinking was approved by the Food and Drugs Administration earlier this year, in the form of "Exubera", a stable, glassy form of insulin. Sadly, the development took more than 15 years to surmount all the obstacles that stand in the way between the "good idea" and the market place. Several of these obstacles have nothing to do with the good idea but result from faulty human relationships. This will continue to be a hard lesson to learn by those in a hurry. As for those of us who were forced into freeze-drying, mainly through economic stresses, the experience has been a fascinating scientific and marketing challenge, in parts even enjoyable, and it has also resulted in lasting friendships across the world. I consider it as 25 years well spent.

Felix Franks
London

Contents

Chapter 1	**Historical Background**	1
	1.1 History of Drying as a Preservation Method	1
	1.2 Advent of Industrial Freeze-Drying	3
	1.3 Elements of Stability	6
	1.4 Why Freeze-Dry?	10
	1.5 Stability: The Downside	12
Chapter 2	**The Process Sequence in Summary**	13
	2.1 Equipment	13
	2.2 Drying Process: Coupled Heat and Mass Transfer	16
	2.3 How Dry Is "Dry"?	17
	2.4 Why Not Freeze-Dry?	18
Chapter 3	**Essential Product and Process Parameters in Summary**	20
	3.1 General Considerations	20
	3.2 Formulation Parameters	21
	3.3 Process Parameters	24
	3.4 Multidisciplinary Nature of Freeze-Drying	24
	3.5 Conclusions	27
Chapter 4	**Essential Physics of Low Temperature and Freezing**	28
	4.1 Chill Versus Freezing: Undercooling	28
	4.2 Deep Chill: Undercooled Water	31
	4.3 Physical Properties of Undercooled Water	32

4.4 Ice: Its Nucleation in Undercooled Water 33
4.5 Ice: Crystallisation and Crystal Habits 42
4.6 Nucleation and Crystallisation of Ice and Solutes in Aqueous Solutions 43
4.7 Polymorphic Transitions, Hydrates and Transient Hydrates 48
4.8 Vitrification as Avoidance of Crystallisation 50
4.9 Freeze-Drying of Complex Biological Materials 52

Chapter 5 Essential Chemistry and Biochemistry Associated with Low Temperature and Freezing 54
5.1 Physicochemical Properties of Water-Based Systems under Conditions of Deep Chill 54
5.2 Cold Inactivation of Proteins 55
5.3 Changing Phase Relationships during Freezing 57
5.4 Eutectic Crystallisation of pH Buffer Components 60
5.5 Effects of Freeze-Concentration on Reaction Kinetics 64
5.6 Complex Ternary and Multicomponent Phase Behaviour 67
5.7 Supersaturation and Vitrification: The Importance of the State Diagram 71
5.8 Aqueous Glasses as Solid Solutions 74

Chapter 6 Physical Properties of Crystalline and Amorphous Solids 78
6.1 Crystalline and Amorphous Solids: A Comparison 78
6.2 Non-Equilibrium Processes in Amorphous Solids 82
6.3 Slow Relaxation during Cooling and Heating 84
6.4 Glass Transition: A Summary 85
6.5 Amorphous States and Freezing Behaviour 87
6.6 Materials Science of the Glass Transition: Its Relevance to Freeze-Drying 88

Chapter 7 Excipients: Their Role in Rational Formulation Design 89
7.1 Definitions and Classifications 89
7.2 Attributes and Requirements 90
7.3 Estimation of T_g of Complex Mixtures 95
7.4 Effects of Excipients on Product Attributes 96
7.5 Implications for Freeze-Drying 97
7.6 Devitrification of Excipients in Dried Amorphous Preparations 99
7.7 Pharmaceutical Glasses: Formulation Aspects 100
7.8 Constraints 101
7.9 Operational Degrees of Freedom 103

Chapter 8 Primary Drying: The Sublimation of Ice 105
8.1 Principles of Coupled Heat and Mass Transfer 105
8.2 Model 105
8.3 Heat Transfer Mechanisms 107
8.4 Influences of Formulation and Freezing Procedure on Ice Sublimation 112
8.5 Interplay of Pressure, Temperature and Time 114
8.6 Adjustable Parameters: Formulation, Concentration and Fill Depth 116
8.7 Ideal Conditions 117
8.8 Accelerated Freeze-Drying 117
8.9 Monitoring Ice Sublimation 119
8.10 Summary 120

Chapter 9 Secondary Drying: The Removal of Unfrozen Water 121
9.1 Unfrozen (Residual) Water 121
9.2 Removal of Unfrozen Water by Diffusion 123
9.3 Effects of Pressure 123
9.4 Effect of Water Content 123
9.5 Effects of Fill Depth and Cake Morphology 124
9.6 Effect of Temperature 125
9.7 Softening and Collapse 128
9.8 Operational Drying Protocols: Temperature Ramping and Stepwise Heating 128

Chapter 10 Some Practical Aspects 130
10.1 Scale-Up and ICH Manufacture 130
10.2 Economics 131
10.3 Vial Closure and Packaging 134
10.4 Process Monitoring and Control 135
10.5 Stability Issues 139

Chapter 11 The Dried Product 141
11.1 Physical State: Crystalline or Amorphous? 141
11.2 Essential Materials Science of Crystalline and Amorphous Products 144
11.3 Essentials of Thermoanalytical Techniques 151
11.4 Fragility and Its Estimation 154
11.5 Monitoring Physical and Chemical Stability 158
11.6 Estimation of Useful Shelf Life 159
11.7 Dynamics in Supersaturated Solutions 161
11.8 Residual Water and Its Estimation 164

11.9 Residual Water: Its Influence on Stability 168
11.10 Unresolved Issues 168
11.11 Reconstitution at Point of Administration 170

Chapter 12 Case Histories 171
12.1 Introduction 171
12.2 Choice of Formulation 173
12.3 Process Cycle Optimisation 173
12.4 Concurrent Product and Process Refinement 173
12.5 Inadvertent Failures: "Unexpected Problems" 175
12.6 Unfamiliarity with Freeze-Drying Parameters 178
12.7 *In situ* Water Release during Storage 179
12.8 Summary and Conclusions 180

Chapter 13 Beyond Freeze-Drying 182
13.1 Alternative Water Removal Technologies 182
13.2 Evaporative Drying: Principles and Practice 183
13.3 Evaporative Drying: The Current State 184
13.4 Comparisons with Freeze-Drying 185

Acknowledgements 187
Bibliography 190
Subject Index 201

CHAPTER 1

Historical Background

1.1 History of Drying as a Preservation Method

The recognition that all organic matter on this planet is either water soluble, or at least water sensitive, can be taken as a starting point of almost everything that is discussed in this book. Following on from this bald statement is the unfortunate fact that aqueous solutions of many organic substances, as well as water-sensitive solids containing even relatively low amounts of absorbed water, tend to be subject to chemical, microbial and/or physical changes, most of which are perceived to be undesirable, perhaps even hazardous or lethal. This realisation goes back many centuries: sailors who travelled on exploratory journeys to faraway places learned early on to pickle their meat and other food supplies in concentrated brine solution, so as to inhibit the growth and proliferation of toxic principles that were known to lead to painful or even incurable diseases.

In more modern times, refrigeration and frozen storage have superseded the ancient pickling processes and other, more refined methods have been devised to render foods stable, safe and acceptable for extended periods. Among such methods, drying takes pride of place, insofar as it does not reduce the perceived eating quality of a product. The advent of "instantisation" technology has led to a wide range of products that bear witness to the successes achieved by controlled drying of aqueous solutions, or of solids that contain appreciable amounts of water. Where desirable, solid states so obtained can be readily reconstituted with only minor losses in eating quality. What is not universally appreciated is that freezing as such, is in fact a drying process. As liquid water is converted into crystalline ice from the original aqueous solution or solid substrate, the residue becomes increasingly dehydrated. Upon rewarming, the ice melts and, ideally, the dehydrated product can be reconstituted into its original state, undamaged by the dehydration/rehydration cycle.

The discovery of freezing as a means of drying was chronicled in detail by one Gerrit de Veer, who was born in Amsterdam some time between 1564 and 1577. As a young man, he accompanied the famous seafarers Willem Barents and Jacob van Heemskerk on two of their journeys in 1595 and 1596. Although his job appears to have been that of an ordinary seaman, with little seafaring experience, he kept a detailed log of the expeditions.[†] Barents' and van Heemskerk's aim was to find a navigable passage around the north of the Asian continent, all the way to China. Barents made several such attempts, all of them unsuccessful, and he died in the Arctic during the 1596 expedition.

According to de Veer's log,[1] a supply base had been established in 1595 on one of the islands close to "Nova Zembla" in the Arctic Ocean, where food and drink were stored for the following year's expedition – the one that proved to become Barents' final attempt to reach China and from which he did not return alive. When the 1596 expedition reached the supply base, the sailors were surprised to find that the beer barrels, which had been left there from the previous year, had not only split open, but that they contained large quantities of ice which, upon thawing, did not taste like beer. At the bottom of the barrels, they found a creamy, unfrozen substance that tasted of yeast and proved to be quite intoxicating. The seafarers had in fact stumbled on the process that is nowadays referred to as "freeze-concentration" and which forms the first stage of freeze-drying and of frozen storage in general.

At about the same time, a more targeted, almost hi-tech application of freeze-drying was already being practised by some South American Indians, who were able to convert several selected potato varieties into "white chuño". This was the name given to a product that does not contain bitter glycoalkaloids, is light, easy to transport and can be stored for extended periods. To produce freeze-dried chuño, the Andean Indians froze potato tubers overnight and then warmed them in the sun, but without direct exposure to the sun's rays. In this manner, the ice was removed by sublimation. Next they trampled the semi-dried product to slough off the skins and squeeze out residual water. Finally the tubers were soaked in cold water for several weeks and sun-dried for a week. The chuño, so obtained, formed a white crust, possessing the desirable

[†] The original Dutch text was first published in 1598. Translations into English, French and Latin were published in the same year. A further English translation, by Phipps, was published in 1609, to be followed by yet another translation, by Purchas, introduced as follows: "Reader, I present thee here three admirable Voyages of Discovery made by the Dutch, no whit envying their due prayse, but honouring their worthy Acts and Arts".

storage properties of a freeze-dried, stable potato product, which was first described by chroniclers who accompanied the Spanish invaders of the Inca Kingdom.[2]

1.2 Advent of Industrial Freeze-Drying

In reality, the consequences of freezing and thawing are usually more complex than those described by de Veer, or as performed by the South American Indians. For example, the removal of water from complex solid substrates, such as fish or meat, is accompanied by an irreversible aggregation of muscle tissue, which is then unable to reintegrate enough water into the muscle tissues after thawing, giving rise to what is known as "drip loss". Other deleterious processes that may occur during drying include the rupture of cell membranes (*e.g.* soft fruit) or enhanced lipid oxidation, leading to off-flavour and rancidity (*e.g.* fish). Nevertheless, the controlled drying of food, leading either to the so-called intermediate- or low-moisture products or to "instantisable" products has made great strides and has established itself as an important branch of food process development. Although frozen distribution and storage of food is still widely practised, freeze-drying, *i.e.* the sublimation of ice from a previously frozen product, has been largely superseded by other, more economical drying methods. Its current use is limited mainly to the production of high-value products, *e.g.* Japanese fish specialities, or products where losses of volatiles would seriously impair quality. The place of freezing and freeze-drying in food process technology, as it existed in the 1970s and how it was predicted to change during the 1980s, is well summarised by the contributions presented at a Royal Society Discussion in 1974.[3] With the benefit of hindsight, the reader is left to decide to what extent the predictions actually came true.

The situation is totally different in the pharmaceutical industry, where freeze-drying has gradually established itself as the standard method for rendering aqueous (or even some non-aqueous) solutions of bioactive substances into solid, stable states. An important difference between food and drug processing lies in the contribution of processing towards the total cost. In the case of food production, the various processing stages account for an appreciable contribution towards the total cost, whereas in the pharmaceutical industry the cost of downstream processing, although appreciable, is usually only a minor factor, compared to the cost of the purified bioactive raw material, the drug substance. Another reason is that the regulatory requirements in the food industry are less stringent than those which govern purity, sterility and stability criteria, applied to pharmaceutical preparations, and in particular to parenteral products.

Yet another important difference between dried food products and pharmaceutical preparations is that the former are usually sold by weight, almost irrespective of quality, whereas the value of the latter is measured in the so-called international units of activity (IU). Thus, for enzymes, the IU measures the amount of substrate that is turned over in unit time by unit weight of enzyme, at a given temperature. Hence, freeze-dried coffee, even when of a somewhat indifferent quality, will find its way onto the supermarket shelf, where it can be labelled and sold as "coffee". The same is not true for, say, an industrial or a therapeutic enzyme. Any losses in specific enzyme activity, incurred during downstream processing, including the final freeze-drying operation, downgrade the value of the finished product, since this is not measured as kg (or mg) of enzyme, but as IU of enzyme activity per kg (or mg).

The choice of a particular drying method is also partly dictated by the scale of the operation. Where it is necessary to remove vast quantities of water from a very dilute solution, *e.g.* for purposes of waste water purification, freeze-drying is completely unrealistic. On the other hand, the method has been successfully used to dry valuable books and documents in libraries that have suffered flood damage.

Since we are here concerned mainly with pharmaceutical and, especially, biopharmaceutical freeze-drying, processing costs are of secondary importance. Care must however be taken that limited freeze-drying capacity does not give rise to production problems. Thus, where the total drier capacity threatens to become the limiting factor in the production cycle, the best remedy is to examine whether, and how the cycle length might be reduced, commensurate with the maintenance of an acceptable product quality. This is a problem to which we shall return several times.

Probably the first comprehensive monograph on freeze-drying was published in 1949.[4] It traces the early development of the technique, going back as far as 1813, when William Hyde Wollaston, in a lecture to the Royal Society, demonstrated the relation between vapour pressure and temperature, and the cooling effect of evaporation. Wollaston called the procedure "sublimation", which he defined as the process in which a solid (ice) is converted into a gaseous state and then recondensed as a solid, thereby totally avoiding the intervention of a liquid state during the process.

The actual freeze-drying process was first tested and used in 1890 in Leipzig, Germany. It became of practical importance during World War II, when the Canadian Red Cross had to make weekly deliveries of up to

2000 units of human blood plasma. With the help of several US companies, a peak production of 100,000 units per week was eventually achieved. At about the same time, R.I.N. Greaves, working at the University of Cambridge, began the development of more advanced equipment, which was later employed in the first commercial production of antibiotics. During the 1950s, freeze-drying began to be routinely used by the food and drug industries. The rapid development of the technology can best be judged by the number of publications in the field, which grew from 10 in pre-1930, to 350 over the period 1930–1945. When freeze-drying first became an important process technology in the pharmaceutical industry, the number of publications began to climb rapidly, reaching *ca.* 600 in the year 2000. The first patent was issued in 1934, and the number of US-granted patents during the period 1945–2003 exceeds 400; although references to freeze-drying, as a manufacturing process, feature in many more food and pharmaceutical patents, although not always as the inventive step.

It is interesting to chart the importance attached to the various aspects of the technology at different periods. Thus, during the 1970s, emphasis was placed on modelling the process in terms of heat transfer through frozen layers, and the problems of "collapse" received intensive study. During the following decade, emphasis shifted to the development of electronic devices to control the process. Investigations also commenced on additives and "bound water". The freeze-drying of blood derivatives (oxyhaemoglobin, albumin and clotting factors) became of commercial importance. The decade also witnessed the publication of numerous studies by M.J. Pikal (sometimes referred to, and rightly so, as "the king of freeze-drying"), dealing with more advanced and realistic models of heat and mass transfer. A particular strength of Pikal's contributions derives from his often-expressed philosophy that experimental tests need to be applied to validate all theoretical results. During the 1990s, several reports appeared analysing the economics of large-scale batch freeze-drying. This is a subject to which we shall return later. Product formulation issues also received attention, and the significance of amorphous states vis-à-vis crystalline states began to dawn on the pharmaceutical industry, especially when related to biopharmaceutical products that cannot be crystallised. At the same time, regulatory hurdles became of importance, reining back the more adventurous product developers. Although freeze-drying of foods had declined during the second half of the 20th century, the appearance of probiotics and other "nutroceutical" products regenerated interest in the freeze-drying of bacterial cultures, particularly lactic acid bacteria, where it soon became apparent that isolated molecules are easier to stabilise against deterioration than living species. Even more ambitious attempts have included the long-term stabilisation of freeze-dried mammalian blood cells, embryos, spermatozoa and even organs. The informed reader is left to differentiate between success and failure.

The past two decades have also witnessed the publication of several texts on freeze-drying, some of them focused, allegedly, on the food and pharmaceutical industries. They range from single-author monographs[5,6] to edited volumes, with contributions by several authors,[7] and international conference proceedings,[8] which, for various reasons, are least satisfactory, unless carefully edited for linguistic problems. Central topics such as engineering principles of heat and mass transfer, reviews of available equipment, its correct use and maintenance, simple physical principles associated with freezing and drying, and good manufacturing practice and registration issues are more or less adequately addressed. However, the recent realisation of the overriding importance of product formulation and the thermophysical properties of water-soluble amorphous products are largely ignored in the available monographs, although they figure quite largely in the current scientific literature. It is one of the chief aims of this volume to restore the balance and to place the topics of product composition and formulation centre stage, where they rightfully belong. In other words, they are discussed in terms of "Food Polymer Science", a term coined as far back as 1988.[9]

The products that appear in this book fall mainly into the following categories:

- Drugs (conventional and biopharmaceuticals)
- Microbial/yeast cultures
- Mammalian cell cultures
- Blood products
- Human and veterinary vaccines
- Diagnostic constructs
- Research enzymes/nucleotides/lipid and glyco-conjugates
- Industrial enzymes

1.3 Elements of Stability

According to perceived wisdom, the solid, dry state of otherwise labile pharmaceutical materials can be equated with "stability", defined by an extended shelf life, with reference to specified storage conditions. Let us consider the exact meaning of "stability", although even the more basic word "dry", as used, often somewhat carelessly, in freeze-drying parlance, has led to lengthy arguments between judges, advocates and the so-called expert witnesses in courts of law. In the *Oxford Dictionary*, "dry" is defined as "devoid of all natural moisture", or in more technical terms, "anhydrous". However, as mentioned earlier, in our ecosphere, even under low relative humidity conditions, anhydrous organic materials have no existence. We must therefore look for another, more practical definition of "dry".

In purely pharmaceutical terms, the meaning of "stability" is clear: it relates to the maintenance of the exact molecular identity of a given drug substance. Chemical instability can arise by different routes, which include hydrolysis, oxidation, isomerisation, condensation, racemisation and reactions with other components in solution.[10] This subject is discussed in more detail in Chapter 5. A particularly troublesome group of reactions is common in the processed food industry. These non-enzymatic Maillard or "browning" reactions occur between peptides and reducing sugars, especially in semi-dry environments. They are of only minor significance in pharmaceutical processing, and the interested reader is referred to one of the many monographs that deal with this particular problem.[11]

In the vast majority of chemical destabilisation/bioinactivation processes, water acts either as a catalyst or it participates as a reactant and/or product. It therefore seems logical to conclude that the removal of water should eliminate many causes of chemical instability. The situation is not quite so clear however in the case of physical instability. Processes of concern that can take place in the solid state include polymorphic solid/solid transitions and the compaction of powders. Even low levels of water vapour sorption may lead to other undesirable changes, *e.g.* solid/liquid phase separations, recrystallisation in the solid state or polymorphic transitions. Since all these processes occur only in the solid state, it follows that they cannot necessarily be eliminated by drying. The important factors in physical and mechanical stabilisation are the actual state of the solid produced by drying, the level of residual water and the temperature and pressure employed during processing and storage.

Special consideration must be given to biopharmaceuticals, *i.e.* complex preparations containing one or several biopolymers, often of peptide origin, although similar considerations apply to polysaccharides, phospho- and glycolipids, nucleotides and conjugates of the above, *e.g.* immunoglobulins or lipoproteins. Even more sensitive to destabilisation and inactivation are supramolecular assemblies such as viruses, multisubunit enzymes and lipid vesicles. Here, more subtle effects are at work, which tend to destroy biological activity, even without affecting covalent bonds. Thus, a protein relies for its specific biological activity on a unique molecular shape in which one or several polypeptide chains with specific amino acid sequences adopt a specific configuration in which the amino acid residues are held in position by non-covalent, short-range interactions with one another, with ions and with a solvent (water). Apparently the only covalent links, apart from the peptide bonds between amino acids, which contribute to protein stability are disulfide bonds between cysteine residues.

Instability arises when the so-called native (bioactive) protein state loses its unique folded configuration. This process is referred to as "denaturation". It can be brought about by changes in pH, some salts, temperature, pressure and certain organic additives. Here the important point is that, in principle, denaturation should be reversible, with the native state being reconstituted as the original environmental conditions are re-established; with careful experimentation this can often be achieved, even for quite complex structures.[12] Indeed, controlled denaturation, followed by renaturation, forms the basis of several downstream processing and purification protocols.

In practice, extreme care must be taken in the various processing stages, including freeze-drying, if high recovery levels of native, bioactive protein are to be achieved. The main reason for irreversibility is that in their denatured states, proteins become more vulnerable to the various chemical deterioration mechanisms, already briefly discussed. Additional inactivation mechanisms include the formation of intra- or interpeptide bonds (aggregation) and disulfide bond rearrangements, both of which are essentially irreversible. Some common chemical reactions that lead to inactivation are summarised in Table 1. Most of them are pH-sensitive, as shown schematically in Figure 1. In solution,

Table 1 *Common chemical reactions leading to the transient or permanent inactivation of proteins*

Reaction	*Location*	*Notes*
Oxidation	MET	Most susceptible
	CYS	Catalysed by metals
	HIS, TYR, MET, CYS, TRP	Photooxidation, pH-dependent
S–S interchange		Catalysed by –SH
Hydrolysis	Deamidation at ASX, GLX	Buffer concentration effects
	Peptide bond cleavage	ASP-PRO worst case, acid catalysed
Beta-elimination	CYS, SER, THR, PHE, LYS	Accelerated by high pH and metals
Racemisation	MET, ASP	Much faster in proteins than in free amino acids

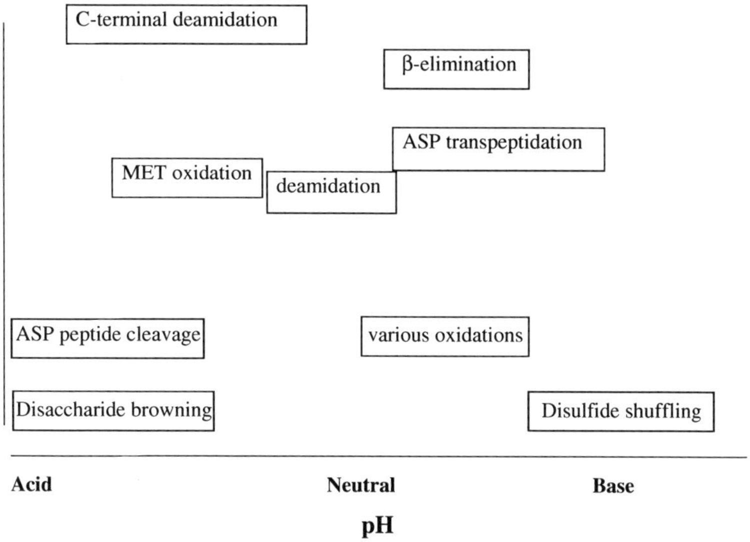

Figure 1 *Susceptibility of protein inactivation reactions to pH*

the native state is considered to be in dynamic equilibrium with several unfolded states, which means that, over time, every molecule samples several states; the degree of occupancy of these states, averaged over time, is given by an equilibrium constant *K*, where

$$K = [\mathrm{N}]/\prod[\mathrm{D}_i] \tag{1}$$

where [N] refers in the concentration of the unique native (bioactive) state and $[\mathrm{D}_i]$ the concentration of the *i*th denatured state. Little is as yet known about the numbers and structures of such D states, let alone their respective concentrations. A caveat needs to be entered here: Protein stability studies are generally performed in dilute solutions. It is questionable whether, or how, the results and conclusions from such studies can be immediately related to reactions in supersaturated, highly viscous solutions, such as are produced during the freezing process, or in the formulated, dried state.

It might be thought that freezing, followed by thawing, constitutes a reversible processing cycle. Although frozen foods are an example of this technology, the frozen product, when reconstituted, rarely matches the original state in quality. The main reason is that freezing (*i.e.* drying) is a process that takes place at very high rates in most substrates and one that can hardly be controlled. Thus, thawing cannot be regarded as a reversal, at least in chemical terms, of the freezing process. The consequences are discussed in more detail in Chapters 5 and 11. It will now be apparent that the freezing (concentration) of a solution must increase the probability of random collisions, which can accelerate aggregation, leading eventually to precipitation. In practice it is more important to know whether, on reconstitution in aqueous solution, and at the point of use, the bioactive state is rapidly regained.

1.4 Why Freeze-Dry?

Having once accepted the consensus view that "dry" is indeed synonymous with "stable", the question arises why the drying process needs to be performed by the circuitous route of freezing, followed by sublimation of ice. On the basis of the total energy balance, a standard process cycle necessitates several cooling and heating processes. When compared to a simple evaporation of water, say at 37°C, freeze-drying as a means of removing water, must be regarded as an extremely wasteful utilisation of energy. There are several additional cost factors to be considered, and the "true" cost of freeze-drying is examined in Chapter 10. The following question may well be posed: why has freeze-drying become the universally preferred method of rendering pharmaceutical products stable? It is interesting that several of the reasons advanced are based on misconceptions, while others rely only on marketing wisdom.

The following summary gives the main responses that have been advanced by seasoned industrial managers when posed the question: why freeze-dry?

- Low temperatures protect the active material during processing.
- Freeze-drying is a process that is approved by regulatory authorities.
- Freeze-drying can be performed under sterile conditions.
- The dried product can be rapidly rehydrated.
- The finished product format (a low-density, porous plug) is recognised and accepted by end users.
- Freeze-drying is associated with quality.
- Freeze-drying has the reputation of being simple to perform.
- It has been successfully used by other companies and/or for other products.
- This type of product has always been freeze-dried.

It is seen that the reasons advanced are indeed a strange mixture. They are listed here in no particular order of significance or veracity. The most important reason appears to be that freeze-drying of pharmaceutical products has long been accepted by those national and international authorities, which regulate the granting of product licences. This alone would be enough to stifle any development of alternative drying processes. Even if technically successful, they would have to be submitted for lengthy and costly examination and validation by the various regulatory authorities.

Factors in favour of freeze-drying are undoubtedly the ability of carrying out the process under conditions of absolute sterility and the rapid reconstitution of the dried product at the point of use. It is also true that the white solid plugs, so characteristic of properly freeze-dried solutions, are widely recognised and accepted as a mark of quality by end users. The most common replies to the question "Why freeze-dry?" are "everybody else does it" or "this product has always been successfully freeze-dried". The widespread conservatism of company production departments is so well documented that it needs no further elaboration.

We now come to three doubtful reasons, in particular the view that low temperatures are beneficial processing conditions. It will be shown in Chapter 5 that such a statement needs careful qualification; it is certainly not a universal truth. As to the quality of the freeze-dried products, here again questions arise. Product catalogues distributed by manufacturers of biochemicals bear witness to apparent doubts about quality, because the need for refrigerated storage is usually specified and, for many biopharmaceutical products, deep frozen storage is apparently an absolute requirement. If this is indeed the case, then the slow removal of ice from the frozen solution by sublimation is surely superfluous. A simple freeze-thaw process would suffice.

The widely accepted, but mistaken, claim that freeze-drying is a simple process to perform has probably been responsible for its almost complete neglect by chemical engineers as a subject worthy of study. Although "*drying*" is one of the accepted generic unit operations that feature in chemical engineering texts, freeze-drying does not even merit a mention, despite its technical complexity and its undoubted commercial importance. As it becomes apparent in subsequent chapters, freeze-drying is anything but simple, if not in concept, then certainly in its practical operation in pharmaceutical manufacturing technology.[170] Its neglect by chemical engineers is remarkable, because in this era of computer simulation, freeze-drying, as usually performed by the pharmaceutical industry, should provide interesting and unique systems for study by computer modelling.

1.5 Stability: The Downside

Usually, and for obvious reasons, stability is discussed in terms of its benefits. Any such discussion would, however, be incomplete without some mention of commercial situations where stability is undesirable, or even dangerous. Enzymes are used in many food manufacturing processes, mainly in the form of lyases (amylases, phosphorylases, lipases, *etc.*). Health and safety considerations demand that they must have been removed entirely before the finished products reach the marketplace. In a similar manner, any extraneous enzymes used as processing aids in protein purification protocols have to be removed prior to freeze-drying.

The situation is even more dangerous for pathogens that can undergo sporulation during drying, *e.g.* canned meat products, that contain minute residual traces of *Clostridium botulinum.* The spores formed during processing possess a very high degree of resistance to any treatments aimed at their destruction. They remain in this dormant and highly resistant state until their germination to a vegetative state takes place. The mechanisms of such germination triggers are not yet well understood, and the incidence of, and fight against, pathogen stability by sporulation continues to present major challenges to the food industry.

CHAPTER 2

The Process Sequence in Summary

2.1 Equipment

Basically, the freeze-drying cycle, as applied to a solution, consists of a sequence of three distinct processes: cooling to below the freezing temperature T_f, in order to maximise the ice content; sublimation of ice at some subfreezing temperature, usually performed under reduced pressure; and finally, the removal of residual, unfrozen water from the solidified solution. The essential equipment thus consists of a refrigeration device, in which the solution can be cooled to the required temperature and which can be evacuated to a low pressure. A further requirement is a means of heating the frozen material, so as to sublime the ice and to remove the residual, unfrozen water, trapped in the solidified product. Finally, a condenser is required to remove and condense the water vapour from the proximity of the solidified product during the drying process. Simple, laboratory-scale equipment usually consists of a bath, containing liquid cryogen in which a flask, containing the solution, is immersed, and where it can be rotated during the cooling process, in order to maximise the cooling rate (film freezing); see Chapter 4. At completion of the freezing stage, the flask is attached to a vacuum manifold, where ice sublimation may be performed, either at ambient environmental temperature or by immersing the flask in a water bath, the temperature of which can be controlled. It is assumed that the latent heat of ice sublimation will maintain the product at a temperature low enough to prevent deleterious changes within the solidified material.

While this procedure, when carefully performed, may be satisfactory for the laboratory-scale freeze-drying of reasonably robust materials, the process, when applied to the production-scale purification of pharmaceutical, and especially biopharmaceutical products, requires major refinements. Where sterility is of overriding importance, the whole procedure, from vial filling to the final sealing stages, must be performed under strictly controlled environmental conditions. It is also important to maintain a good temperature control and to be able to monitor the product temperature during the process. As will presently be discussed, the problems associated with this requirement are by no means trivial. A typical configuration of many pharmaceutical freeze-driers is shown in Figure 1. It consists of a vacuum-tight chamber, fitted with shelves, the temperature of which can be controlled with a circulating heat exchange fluid. The solution to be dried is typically contained in vials or ampoules, resting on the shelves, or occasionally in syringes. Alternatively, as in the case of large-scale manufacture of enzymes or of bacteria, the solution may be poured into open trays that are then placed on the freeze-drier shelves.

The heat exchange system is supplied with a pump, which circulates the fluid through the shelves. The circulation system must be capable of maintaining the shelf temperature at the set desired values.

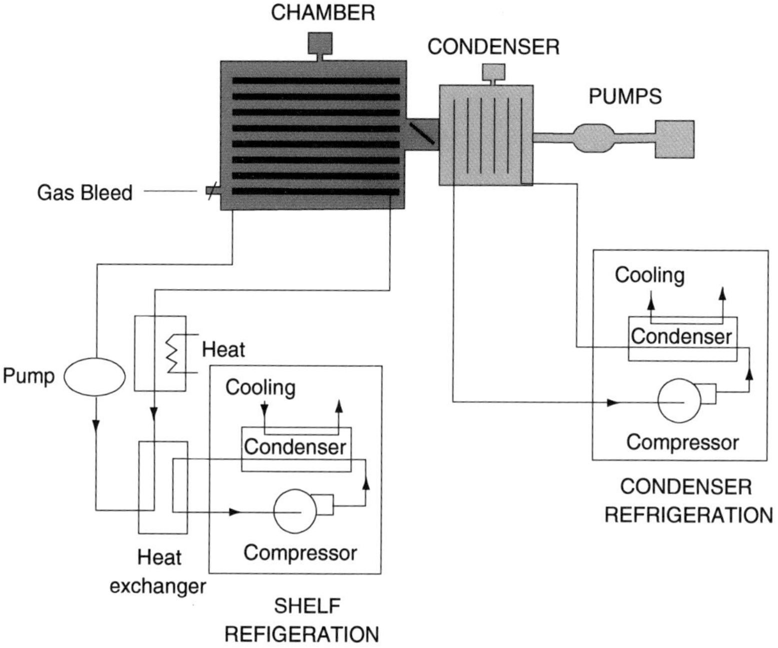

Figure 1 *Schematic of a typical industrial freeze-drying plant, showing the chamber and the adjoining condenser, which can be separated by a valve; each unit is served by its own refrigeration/heating plant. Adapted from M. J. Pikal (unpublished work)*

The vacuum system must be able to maintain the chamber pressure at below the saturation vapour pressure (SVP) of ice at the chosen process temperature. Although it is not universally recognised, freeze-drying can also be performed without the need of a vacuum system. The "laundry model" in which, even on the coldest of days, frozen laundry on a washing line can be air dried, albeit at a low rate, has been cited as an example.

A condenser system, fitted with its own compressor and heat exchange system, serves to remove the water vapour liberated by the subliming ice from the chamber. The system must be able to operate efficiently well below the chosen sublimation temperature and up to high temperatures, so that the accumulated ice can be efficiently removed. It must also be of a capacity sufficient to prevent redeposition of water onto the surface of the drying product.

Modern equipment will also contain computer-based control and monitoring systems by means of which a desired drying cycle programme can be preset. Finally, for pharmaceutical applications, it is essential to prevent cross-contamination between consecutive batches. A "clean-in-place" (CIP) system and a "sterilise-in-place" system are therefore provided by means of which the chamber can be cleaned between successive production cycles.

The whole process can thus be divided into five distinct stages:

(i) The solution, suitably formulated, is frozen on refrigerated shelves.
(ii) The chamber pressure is reduced to the desired value (not necessarily the lowest value that can be achieved) and ice is sublimed by the supply of heat to the product, and the water vapour is removed from the chamber and condensed as ice (*primary drying*).
(iii) The remaining unfrozen water trapped in the product is removed by further warming at a controlled rate (*secondary drying*).
(iv) The containers are sealed and removed from the drier.
(v) The condenser is heated to melt and remove the collected ice.

Once formulation and container details have been decided, the entire process cycle can essentially be controlled by only three parameters – temperature, pressure and time. It is important to remember also that freeze-drying in practice is always a compromise between competing demands and often leads to variable or unpredictable results. The standard industry view used to be, and still is in some quarters, that acceptable freeze-drying conditions can only be determined by trial and error.[13] We hope that this common view will be dispelled in the following chapters. Indeed, it is the main purpose of this book to demonstrate that almost *all* freeze-dried products can be made with high yields, be easy to rehydrate and, depending on the formulation, possess an acceptable shelf life under "room temperature" storage conditions. A state-of-the-art understanding of the interplay between the formulation and the three controllable process variables (shelf temperature, chamber pressure and time) allows the design of efficient processes to produce high-quality products in acceptable yields.

2.2 Drying Process: Coupled Heat and Mass Transfer

The basic process simply requires that sufficient heat is supplied to the product via the shelves to exactly balance the heat absorbed by the sublimation of the ice, *i.e.* heat supplied to the product = heat required for sublimation:

$$K_v(T_s - T_p) = \Delta H_s(dm/dt) \quad (1)$$

where K_v is a heat transfer coefficient, T_s the shelf temperature and T_p the product temperature. ΔH_s is the heat of sublimation of ice at T_p and dm/dt is the rate of mass transfer of water vapour from the chamber to the condenser surface. The significance and estimation of the various quantities in Equation (1) will be discussed in detail in Chapter 8.

Heat transfer occurs by a combination, not necessarily of a linear nature, of three mechanisms:

(i) radiation, mainly from the walls and the door of the freeze-drier;
(ii) conduction by direct contact between solid and/or liquid interfaces, *i.e.* shelf/container and container/solution;
(iii) convection through the gas phase (collisions between gas molecules).

The relative contribution of each mechanism to K_v depends on many factors, also to be discussed in Chapter 8.

Mass transfer (water) depends on

- fill depth (cake thickness);
- total solid content;
- cake porosity (governed mainly by ice crystal size distribution);
- surface area from which sublimation can occur.

All these factors help to determine the resistance to sublimation.

2.3 How Dry Is "Dry"?

When considering the meaning of "dry", our conventional understanding of the term, *i.e.* the ability to transfer liquid water from one surface to another, is of little help. For example, the term "bone dry" is in common usage, although some fresh bones have a surprisingly high water content. An adult femur (without epiphysis) contains *ca.* 22% w/w water, yet to the touch it feels to be no more "wet" than tooth enamel, which contains a meagre 3% w/w of water. Not surprisingly, therefore, misconceptions surround a quantitative description of "dryness", especially in a court of law.

Residual water in freeze-dried preparations forms an important determinant of their stability. This quantity is usually expressed as weight % water in the dried product. A "low" residual water content, so expressed, is usually aimed at, but it obscures, the actual water content in terms of mole water per total number of moles of other components, if only because the molecular weight of water is very low compared to that of most drug molecules and excipients. It is therefore instructive to consider the molecular ratio, rather than the mass ratio, of water:solid, especially for products with components of differing molecular weights. If a 5% w/w solution of a monosaccharide (molecular weight 180) is dried to a residual water content of 5% w/w, then the mole ratio of water/sugar in the dried product is 1:2, *i.e.* the "dry" substance actually contains 28 mol% of water. On the other hand, a 5% w/w solution of a protein with 150 amino acid residues (at 66 kDa), when dried to the same residual water content of 5% w/w, will have a final mol ratio of water/protein of 120:1, equivalent to about one water molecule per amino acid residue. It is left to the opinion of the practitioner whether either or both such products should be considered as "dry". In light of these aspects, it must therefore be concluded that the standard practice of expressing residual water in terms of weight fractions may be convenient, but is grossly misleading, where mixtures of substances differing substantially in their molecular weights are involved.

Conventional wisdom, mainly of an anecdotal nature, has it that the quality of certain products, especially proteins, is adversely affected by "overdrying", although there appears to be no well-founded evidence for such claims. No general agreement exists, regarding what might be an optimum residual water content, to be aimed at during the final stages of drying. The effects, if any, of low levels of residual water on the physical and/or chemical stability of solid products are still subject of intense research activity, as are also the claimed deleterious effects of "overdrying". That long-term chemical stability can indeed be reduced even by low levels of residual water is demonstrated in Figure 2

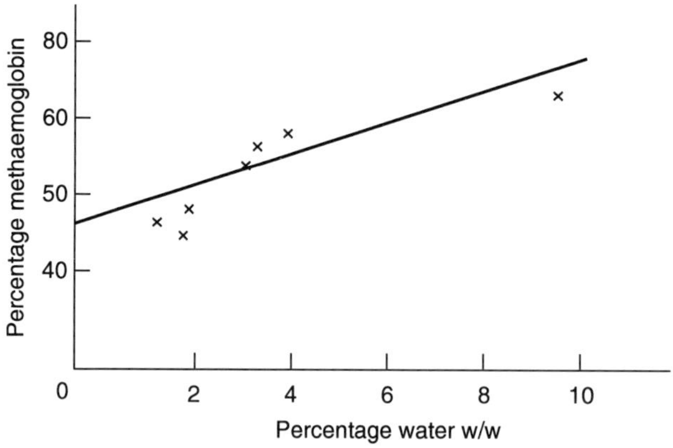

Figure 2 *The effect of residual water on the long-term (4 years) decomposition to methaemoglobin of a 6% solution of freeze-dried haemoglobin, formulated with 0.2 M sucrose and stored at 23°C*

Table 1 *Basic energy requirements for the freeze-drying of 1 kg of an aqueous solution containing 5% solids (sucrose), based on the following assumptions: ice sublimation is performed at −40°C, with a condenser temperature of −60°C (units: kJ kg^{-1})*

Operation	*Theoretical energy needs*	*Refrigeration duty*	*Total energy needs*
Cool/freeze to −40°	500	500[a]	1000
Vacuum pump	2000		2000
Sublimation	2900		2900
Condense water vapour at −60°	3000	7400[b]	10,400
Total			16,300

[a] Coefficient of performance (COP) of refrigerator = 1.0.
[b] COP for water vapour condensation = 0.4.

(M.J. Pikal, personal communication). Bearing this in mind, there are uncertainties as to at what level of residual water the product becomes "overdried". In the experience of the author, the so-called overdrying is not caused by the removal of too much water but by faulty processing conditions, *i.e.* not overdrying, but overheating during drying.

2.4 Why Not Freeze-Dry?

The replies resulting from a poll of process managers in pharmaceutical companies to the question "Why freeze-dry?" were discussed in the previous chapter. One might, however, justifiably ask the question: "Why *not* freeze-dry?" On the face of it, freeze-drying appears to be a very uneconomical process; it is certainly wasteful in energy utilisation. A simple example from the author's archives is shown in Table 1 for a very basic, medium-scale process. The energy requirements shown are limited to those involved in freezing the initial solution and the sublimation of the ice formed, usually $>99\%$ of the initial water content. Not included are the need to remove the ice from the condenser plates or the energy needed to cool and rewarm the freeze-drying chamber and the heat exchange fluid that maintains the programmed shelf temperatures throughout the process. The energy required (16,300 kJ kg^{-1}) compares unfavourably with that required for the evaporation of liquid water at 37°C (2400 kJ kg^{-1}). Other things being equal, therefore, an evaporative drying process appears to be vastly more efficient, both in terms of energy used and the drying period. This issue is further discussed in Chapter 12.

CHAPTER 3

Essential Product and Process Parameters in Summary

3.1 General Considerations

During the process development stage, a number of variables need to be considered; they include

- formulation details, *i.e.* solution composition and concentration
- solution fill volume
- type of container, method of container closure

The physics and chemistry of freezing show many complex and often surprising features, but the industrial freeze-drying process is limited to very few operational degrees of freedom. Thus, for a given solution composition and a given container and closure system, only three process variables exist by which direct control over the drying cycle can be maintained and which, therefore, determine the quality of the final dried product. They are

- shelf temperature
- chamber pressure
- time

On the other hand, the most important factor is of course the product temperature and its change throughout the duration of the process. Its measurement and control are discussed in more detail in several following chapters.

3.2 Formulation Parameters

In the manufacture of a therapeutic product, the initial stage requires the purification of the biologically active component, the drug substance. Purification methods for "conventional" drugs are well established. They are based on synthetic organic chemistry and standard analytical techniques. The situation is more complex for biopharmaceutical products, in particular those based on proteins. The drug substance may then be of animal, plant or microbial origin; it may be obtained from the natural source materials or by recombinant DNA (rDNA) methods. The extraction and purification strategy depends to some extent on the source of the protein. Possible impurities are many, as shown in Scheme 1. The downstream processing always requires several stages and, depending on the source material, may involve a combination of physical and chemical methods. A generalised flow sheet for the isolation and purification of rDNA proteins (r-proteins) is shown in Scheme 2.

The total yield of purified protein is related to the number of stages required and their individual yields. For instance, in an isolation and purification process consisting of n steps, each one of which can be carried out with a 90% efficiency (highly idealised!), the final percentage yield of product is given by

$$\text{Yield (\%)} = 9^n \times 10^{(2-n)}$$

Thus, for a process consisting of five stages, each of which can be performed with a 90% recovery, the yield of the final product cannot

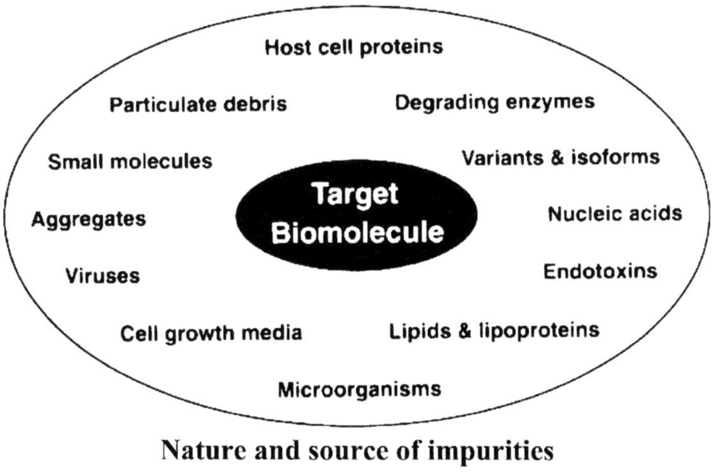

Scheme 1 *Types of impurities commonly found in crude protein extracts and that have to be removed during downstream processing*

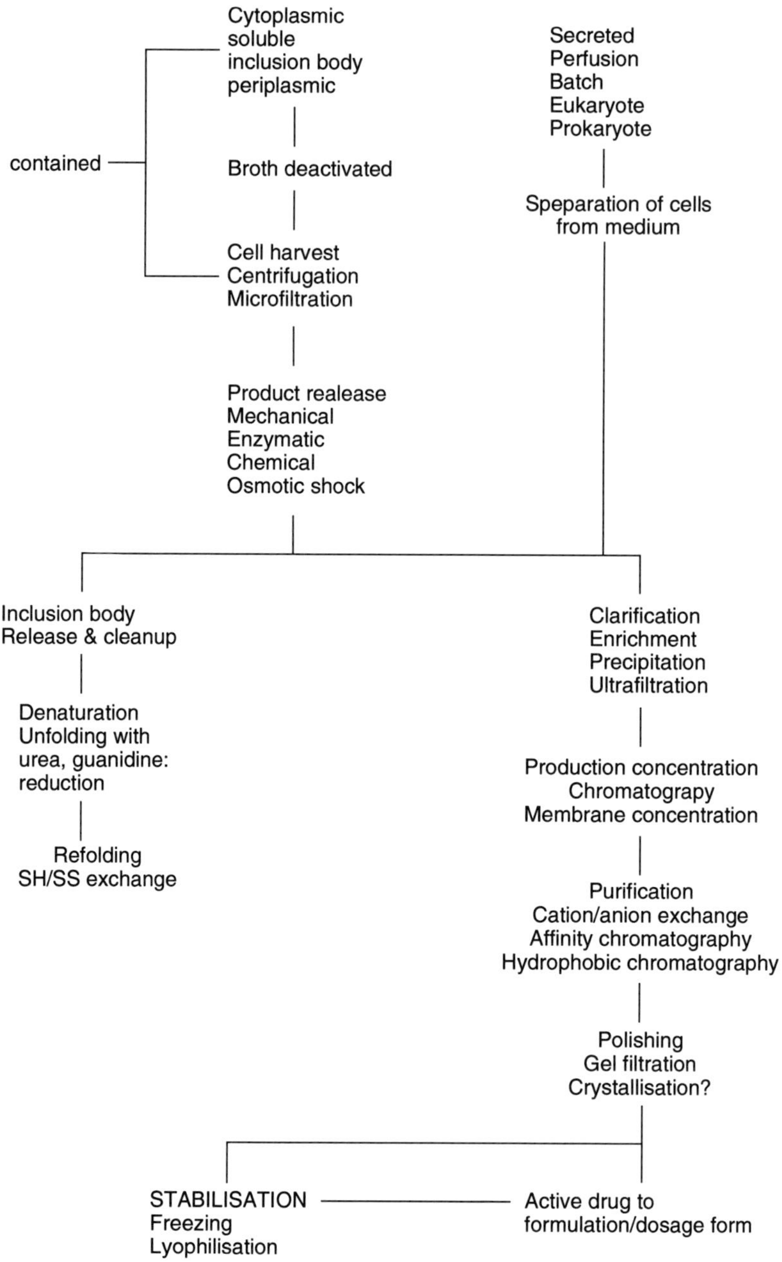

Scheme 2 *Typical downstream process flow sheet for r-proteins*

exceed 60%, which is reduced to 53% for a six-stage purification procedure. Since biopharmaceutical drug substances, whatever their origin, are usually expensive, a loss of 40% during the purification process is a serious factor. A minimisation of the number of stages is therefore aimed at, but it has to be judged against other economic considerations and an acceptable degree of purity.

In the manufacture of proteins for applications other than as therapeutic products, *e.g.* industrial enzymes, purity criteria are not as stringent as they are for parenteral products, in which case acceptable yields may be obtained with fewer fractionation and purification steps. For biochemical and therapeutic uses, however, purity and long-term stability are the overriding requirements. Fractionation, purification and stabilisation by freeze-drying then generally account for 50% of the total production costs, but since such products command high premiums in the market place, production hardly figures in the cost equation. Examples of the typical production cost breakdown for two product/process scenarios are shown in Table 1. The data illustrate that the cost contribution allocated to freeze-drying is almost insignificant, compared to the cost of the purified "raw" drug substance, assumed to be a protein.

Before a purified protein, usually in a dilute aqueous solution, is subjected to the final stabilisation procedure, *e.g.* by freeze-drying, it has to undergo a process of compounding, loosely referred to as "formulation". Very few proteins can survive freeze-drying without the aid of the so-called lyoprotectant additives, which serve to ensure the recovery of full biological activity of the protein at the point of use, whether in the dry solid state or reconstituted in an aqueous medium. The science and technology of lyoprotection, as employed in freeze-drying, is only gradually being put on a quantitative basis; several unresolved mysteries remain (see Chapters 7 and 11).

Apart from lyoprotectants, other substances are also commonly added to the protein solution prior to drying. Some may also have been

Table 1 *Economics of freeze-drying: production cost breakdown (excluding labour) for two typical biopharmaceuticals*

Cost components	*Product I: % cost Small vials, 24-h cycle*	*Product II: % cost Large vials, 4-day cycle*
Pure drug substance	73	65
Solution manufacture	7	10
Containers	3	3
QC assays	8	7

Source: M.J. Pikal (unpublished data).

carried over in the solution during downstream processing. They may include pH buffers, surfactants and salts. A common additive, phosphate-buffered saline solution (PBS) ensures the isotonicity of a reconstituted parenteral product in water, prior to injection. Although from a pharmacokinetic standpoint some of the additives in common use may be innocuous or even beneficial, their presence in the solution may increase technical problems that can be encountered during freeze-drying. For instance, the presence of PBS will inevitably require a longer freeze-drying cycle than would be required for the same solution in the absence of salts. The omission of phosphates and NaCl from the initial solution is therefore helpful, since it allows for a shorter drying cycle. On the other hand, reconstitution must then be performed with PBS solution, rather than with sterile water ("water for injection"), apparently a distinct disadvantage from a marketing standpoint.

3.3 Process Parameters

As already pointed out, the primary process parameters by means of which drying can be directly controlled are shelf temperature, chamber pressure and time. In principle, the condenser temperature can also be adjusted, but in practice it is usually set to the lowest possible value, so as to maximise the ice sublimation rate. Since it is the temperature of the material that determines the drying rate, the control over the process can only be indirectly maintained, usually by coupled adjustments, either continuous or ramped, of the shelf temperature and/or the chamber pressure throughout the drying cycle.

Once the process cycle conditions have been decided upon and the vials have been loaded, the process is automatically controlled by a computer. The freeze-drier performance is monitored, and a sample of a typical output sheet is shown in Figure 1. It records the time course of the production cycle (48 h in this case) in terms of shelf, condenser and product temperatures and chamber pressure. As mentioned earlier, the only *directly* adjustable parameters are the shelf and condenser temperatures, and the chamber pressure. We shall defer a detailed discussion of the information contained in the chart to several later chapters.

3.4 Multidisciplinary Nature of Freeze-Drying

As will become clearer in the following chapters, the successful application of freeze-drying is based on a complex interplay of several

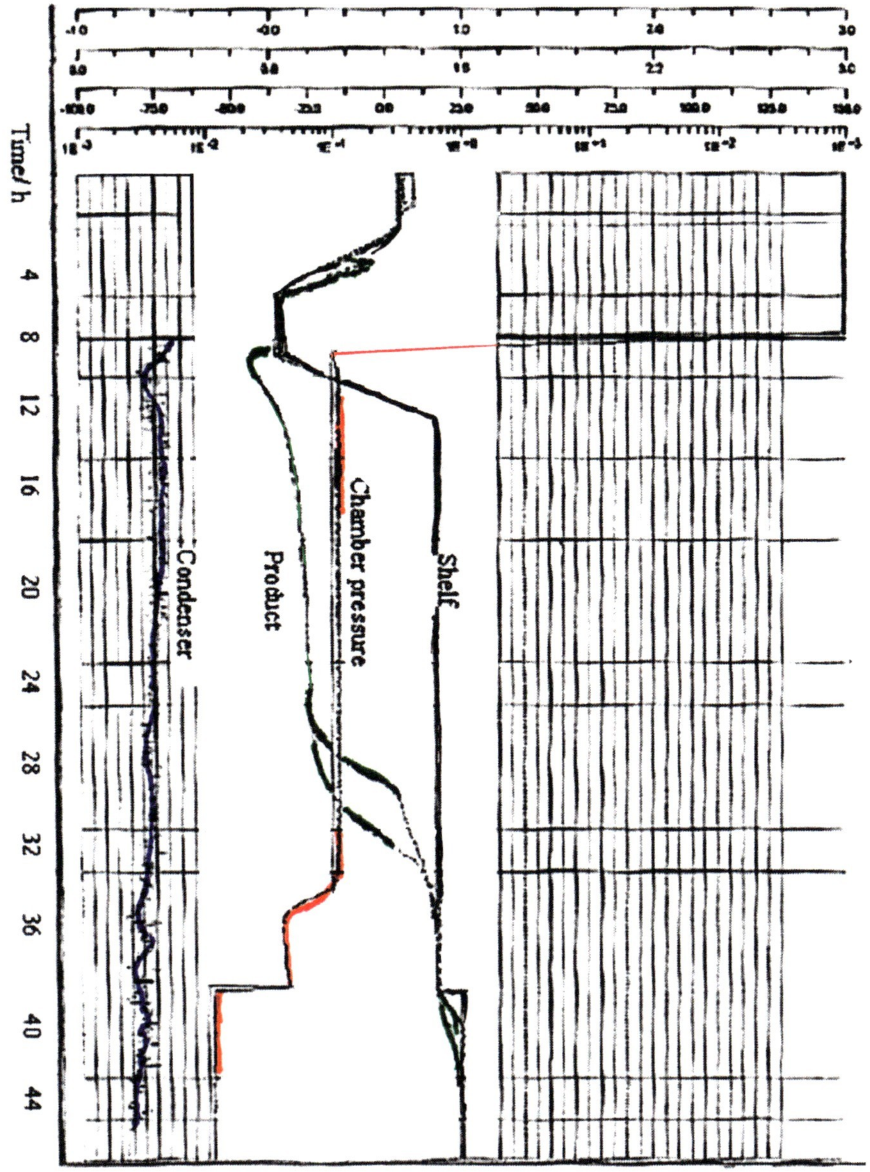

Figure 1 *Freeze-drier recorder output of condenser temperature (bottom, blue), shelf temperature (black), product temperature (green) and chamber pressure (red). Ordinate scales (from top to bottom): 1 and 2 = bar, 3 = °C, 4 = mbar*

scientific principles, some of which are still very poorly explored and hardly appear in university undergraduate curricula. In the author's experience, one rarely finds managers or technologists with an overall broad view of the various stages of freeze-drying. Thus, the trained chemist's experience differs from that of the physicist, the materials scientist or the chemical engineer. And yet, an overall appreciation of how these disciplines interact is of the utmost importance. In addition, experts in the various disciplines within a company will usually be found in different divisions of a line management structure, a custom not conducive to the fostering of the necessary collaboration.

This book aims to highlight the importance of a multidisciplinary approach to freeze-drying. Nonetheless it has been found necessary, although not really desirable, to separate physical, chemical, economic and engineering and other aspects into different chapters. The following paragraph is intended to provide a summary of the important disciplines, some knowledge of which is essential for the development of a suitably formulated product, coupled with a sensible process cycle. The terms printed in italics are those that are of importance and require understanding by the practitioner.

Pharmacokinetics deals with all aspects of the therapeutic effect of the chosen drug substance, suitably formulated, on its target. It forms the very basis of the formulation and processing stages. Of second highest importance is the formulation stage. In the case of biopharmaceuticals based on proteins, some experience of *protein biochemistry and technology* is absolutely necessary. The *formulation* process aims to protect the bioactive substance from the severe stresses to which it will be exposed during any drying process, including those directly due to concentration increases during freezing. The freezing process itself is governed by *ice nucleation* and *crystal growth rates* in supersaturated solutions. It is accompanied by freeze concentration of the solution, a process that may result in both *crystalline or amorphous solids*, or in mixtures. A correct identification of the physical and chemical identities of the product during and after processing requires an understanding of the analytical techniques available for obtaining the necessary information, also for an eventual submission to the relevant regulatory body.

The ice sublimation stage depends on the balance of *heat and mass transfer*. Secondary drying depends on the *diffusion rate* of water through the solid matrix. The long-term stability is governed by the *materials science of crystalline and amorphous solids*, and it is measured with the aid of an array of *physical analytical techniques,* such as *thermal analysis, X-ray diffraction, spectroscopy and chromatography*. For registration purposes, all process and analytical data must be submitted in the required formats, with *product identity, purity, safety, stability and shelf life* playing major roles in the submission for regulatory approval.

3.5 Conclusions

In small teams where good communication between individual staff members is fostered by imaginative management, it is possible for the necessary knowledge and experience to be passed on and become common knowledge, so that high-quality freeze-drying can be performed. The complex nature of biopharmaceutical freeze-drying makes it unlikely that this can be achieved in larger institutions, because most line management structures inhibit the flow of information between groups devoted to formulation and those whose objective it is to devise economical drying cycles. This was brought home to the author when his consultancy work took him to a German company that undertook the freeze-drying of human blood products on a large scale, but with variable results. During a round-table discussion, it became apparent that those pharmaceutical chemists who were responsible for formulation development had never met or spoken to their engineering colleagues, who were responsible for the freeze-drying process. Indeed, for "security" reasons, the latter were not informed of the nature of the products that were to be freeze-dried. This was a scenario that our technologists also encountered in a number of other mega-pharma companies. Usually, even the mention of "glass transition" or its measured value met with blank stares. The complexities were not appreciated, the freeze-drying operation being regarded as a push-button affair, hardly a recipe for success.

CHAPTER 4

Essential Physics of Low Temperature and Freezing

4.1 Chill versus Freezing: Undercooling

At the outset, it must be emphasised that the definition of "low temperature" is arbitrary. A life scientist would normally equate "low temperature" with "subzero Celsius temperature", whereas to a physicist, perhaps studying superconductivity phenomenon, the term signifies temperatures in the neighbourhood of 1 Kelvin. The reason why low temperature is often equated with freezing is probably because the freezing point of "ordinary" water lies near the centre (measured in degree Celsius) of the temperature range -40 to $+40$°C, which we associate with life processes on this planet.

What is not universally appreciated are the fundamental differences between effects of low temperature *per se* and freezing. When we speak of the effects of *low temperature* on liquid water or aqueous solutions, we should consider only changes in various physical and/or chemical properties observed upon cooling a *homogeneous* liquid system, but in the absence of any liquid/solid phase transitions. *Freezing* a homogeneous solution, on the other hand, begins with just such a transition (liquid water → ice). On further cooling, the quantity of ice in the solution increases, as illustrated in Figure 1a for the simplest type of conventional solid–liquid phase diagrams, in this case the binary KCl–water system. As water is removed by freezing, the concentration of the original solution increases with decreasing temperature, a process referred to as "freeze-concentration" in freeze-drying terminology. Figure 1b is a photomicrograph of KCl crystals left after the aqueous solution had been freeze-dried, *i.e.* the ice removed by sublimation. The site of the initial nucleus is indicated in the centre of the picture. The radial ice

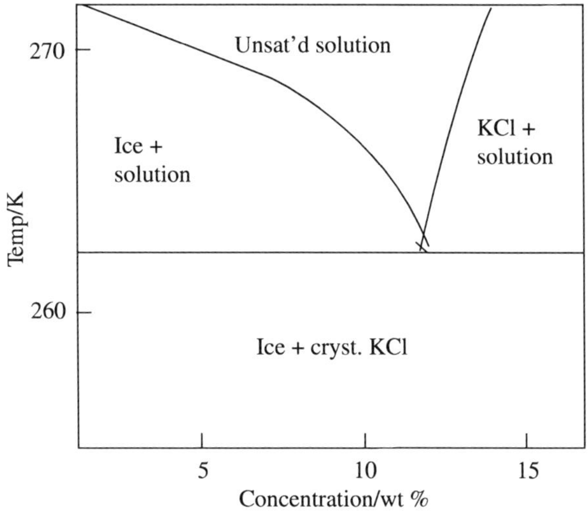

Figure 1a *Solid–liquid phase diagram for the system water–KCl, showing the eutectic point at 262 K. This is the only point at which equilibrium exists between the three phases: ice (crystal), KCl (crystal) and saturated solution*

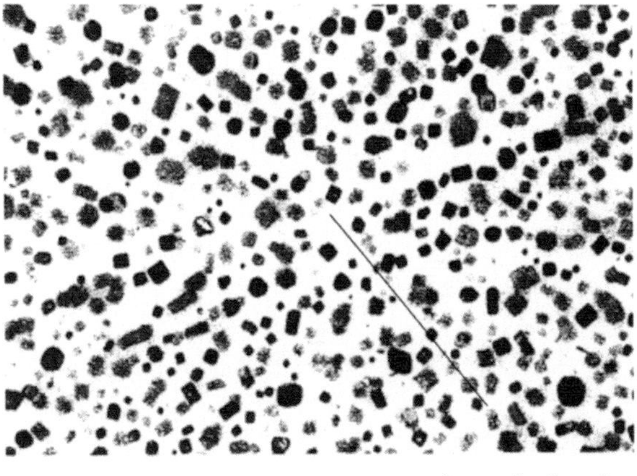

Figure 1b *Electron micrograph of a 20% KCl solution after rapid quenching to −150°C and freeze-drying at −70°C. Note the radial crystallisation pattern, originating from the nucleation site at the centre of the picture. Modified from MacKenzie*[42]

propagation from this point gave rise to the final positions of the KCl crystals.

The fundamental difference between chill and freezing, both *in vivo* and *in vitro*, is illustrated in Figures 2 and 3, respectively. In Figure 2, the metabolic rate of a silk worm larva is monitored by its oxygen uptake, which decreases with decreasing temperature but shows no discontinuity

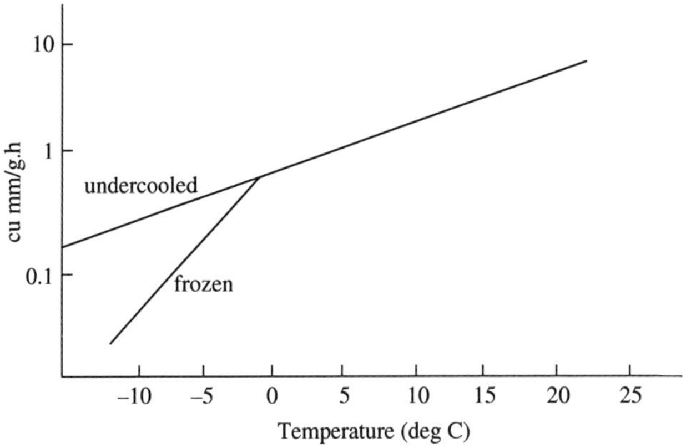

Figure 2 *Oxygen uptake by a diapausing silk moth pupa (Hyalophora cecropia) in the chilled (undercooled) and frozen states, showing the injurious effects of freezing*

at or near 0°C. The indications are therefore that the haemolymph does not freeze but that chill reduces the insect's metabolic activity, as would be expected. The process is fully reversible, so that upon rewarming, the oxygen uptake increases and the insect recovers from a quasi-dormant state. If, however, the larva is held at or below 0°C and touched with an ice crystal, the rate of oxygen uptake exhibits a sharp discontinuity, dropping rapidly upon cooling. The haemolymph has frozen and normal oxygen uptake is not recovered upon thawing, *i.e.* freezing has caused the death of the insect. *Thus, chill causes a slowdown of metabolic processes, enabling the insect to survive winter, whereas freezing is lethal.*

The same conclusion can be reached from *in vitro* studies, illustrated in Figure 3 for the temperature dependence of the specific activity of isolated lactate dehydrogenase (LDH) enzyme in a dilute aqueous solution at pH 7.[14] When stored at room temperature, the decrease in specific enzyme activity shows a typically biphasic behaviour, dropping rapidly by 80% during the first week and more slowly thereafter. Storage in a refrigerator at 4°C (chill) reduces the rate of activity loss, but when exposed to deep chill but unfrozen conditions, say –20°C, the activity loss is practically inhibited. The kinetics of inactivation exactly follow the Arrhenius equation. Within the normal experimental error for enzyme assays, LDH has been recovered at close to 100% of its original activity, even after 6 years of storage. By comparison, *freezing* the enzyme solution at –20°C leads to an almost immediate and irreversible loss of 80% of its activity, with a 100% loss observed after 1 week of frozen storage. Here again, therefore, chill is seen to stabilise, while freezing causes an irrecoverable loss of biochemical activity.

4.2 Deep Chill: Undercooled Water

It has long been known that water, especially when purified to some extent, does not freeze spontaneously when cooled to 0°C. Instead, the liquid state is maintained at subzero temperatures, until at some arbitrary temperature, freezing occurs spontaneously and abruptly. Although the undercooled state may not appear to have any relevance to freeze-drying, because chill depends on the avoidance or prevention of freezing, it is nevertheless important because some degree of undercooling, often down to *ca.* −15°C, can never be avoided during an industrial freezing process, especially one carried out in highly purified water. That undercooling can have useful, even beneficial results, becomes apparent from an inspection of Figure 3, where it is shown that enzyme activity in solution can be completely maintained for long periods in an undercooled aqueous solvent medium, without the necessity for any cryoprotectant additive addition.

The usefulness of undercooled water as a reaction medium has also been demonstrated in biochemical studies. Thus, the kinetics of biochemical reaction sequences can be sufficiently slowed down to become time resolved at subzero temperatures in the unfrozen state. Even more importantly, complex reaction pathways (mechanisms) are unaffected by undercooling, whereas the use of conventional cryoprotectants (glycerol, ethane diol, dimethylsulfoxide, *etc.*) alters the pathways, although not necessarily the nature of the end products.[15] Finally, single cells or even cell clusters can be stored and kept intact for considerable periods in undercooled aqueous media.[16]

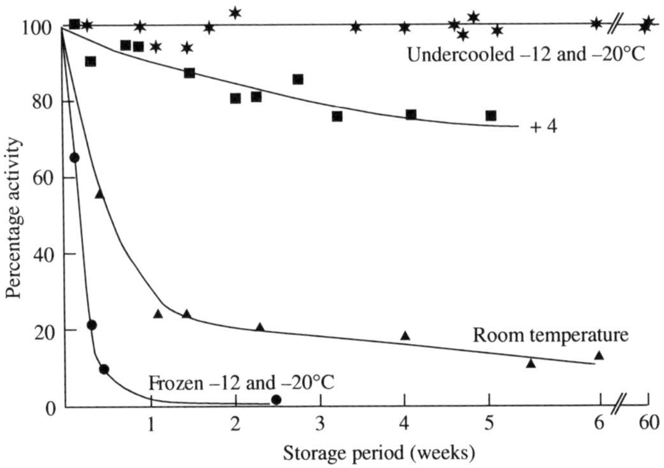

Figure 3 *Maintenance of enzyme activity by LDH, stored under differing conditions. LDH concentration: 10 μg ml^{-1} in phosphate buffer (pH 7). Reproduced from Hatley et al.,[14] with permission by Elsevier*

4.3 Physical Properties of Undercooled Water

All physical properties of liquid water become increasingly sensitive to temperature changes below 0°C.[17] This is illustrated in Figure 4. Most properties appear to diverge at −40°C, *i.e.* they approach infinity, a phenomenon that is, even now, only imperfectly understood and subject to considerable speculation. Ionisation equilibria of water (pK_w) are particularly strongly affected by temperature changes. Thus, at −20°C, pK_w = 16, *i.e.* a neutral pH corresponds to 8, demonstrating that the degree of ionisation of water has been reduced by an order of magnitude compared to its "normal" value of 7 at 25°C.[17] Since dissociation constants of acids (pK_a) and bases (pK_b) are referred to the ionisation of water, it is obvious that all pK_a and pK_b values must also be affected, more or less markedly, by temperature. This is shown in Figure 5 for several acids, where changes in pK_a are plotted against temperature, normalised to 30°C. The curvilinear nature of pK_a(T) plots is common to all acids, although the positions of the minima occur at different temperatures. Changes in pH during cooling are rarely allowed for in the selection of buffer systems during the formulation of solutions destined

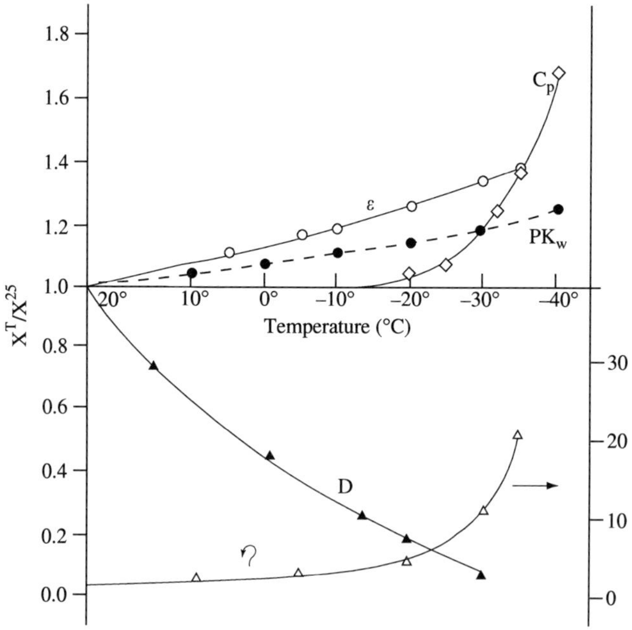

Figure 4 *Temperature dependence of some physical properties of water (X^T) normalised to 25°C (X^{25}): pK_w, dielectric constant (ε), isobaric specific heat (C_p), self-diffusion coefficient (D) – left ordinate and viscosity (η) – right ordinate. Reproduced from Franks[17] with permission from Cambridge University Press*

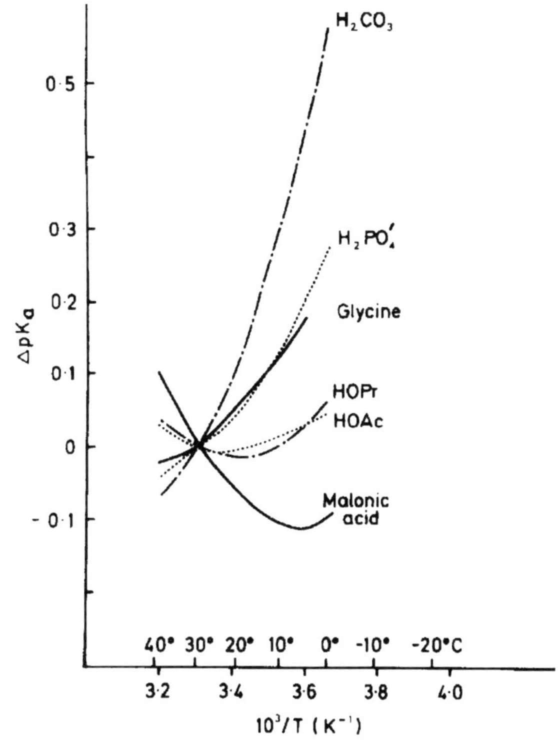

Figure 5 *The effect of temperature on pK_a of some acids used in the preparation of pH buffers, normalised to 30°C. Reproduced from Franks,[17] with permission from Cambridge University Press*

for freeze-drying. As shown, carbonate-type buffers are particularly sensitive to subzero temperatures, whereas acetates show little change in pH in the temperature range of relevance to freeze-drying.

As also shown in Figure 4, the transport properties of water display a strongly non-Arrhenius behaviour at subzero temperatures. Thus, the viscosity of water, when extrapolated to −90°C, is of the same order as that of glycerol at room temperature. This rapid viscosity rise at low temperatures plays a role in limiting the cooling rates that can be achieved in practice.

4.4 Ice: Its Nucleation in Undercooled Water

The occurrence of a phase transition requires a prior nucleation of the daughter phase within the mother phase. Thus, the growth of crystals in a saturated solution requires the prior generation of nuclei, composed of a certain number of molecules, probably in a particular configuration, that can promote the condensation of other molecules, to form a crystal with its characteristically ordered structure. The actual structure of a cluster, capable of initiating crystallisation, is still a subject of uncertainty.[18] As regards the generation of ice in pure liquid water, nuclei arise in the body of the liquid through random density fluctuations, and nucleation kinetics can be reasonably well treated in terms of the growth and decay rates of such molecular clusters and their ability to promote crystal growth. The fluctuations in density originate from self-diffusion (Brownian motion) of molecules, and the probability of nucleation depends on the sizes and structures of high-density domains and their lifetimes. In passing, we emphasise that nucleation of ice is of considerable importance in ecological contexts, *e.g.* cloud physics at one extreme, and freeze tolerance and avoidance mechanisms in living species at the other.[18] It is beyond the scope of this book to discuss nucleation theory in detail, but it is instructive and relevant to industrial freezing processes to summarise some of its quantitative aspects.

Most theoretical treatments of nucleation and crystal growth are based on a model of stepwise addition of molecules to an embryo, up to a critical size, at which the properties of the embryo (*e.g.* its surface free energy) are equated with those of the known solid phase, *i.e.* the crystal.[19, 20]

In the case of water and ice, such a model might be realistic in treating the nucleation of liquid or solid from supersaturated vapour, as it occurs in the upper atmosphere. It seems doubtful, on the other hand, whether the model of stepwise molecular condensation can realistically be applied to the liquid → solid transition, because liquid water itself is already extensively associated and exists as a three-dimensional distorted network of hydrogen-bonded molecules, not too dissimilar from ice. Any mechanistic model of ice nucleation, based on the condensation of individual water molecules onto clusters of molecules with the properties of ice, should therefore be treated with caution.

Ice nucleation is of enormous practical importance, and we therefore summarise the basic features of the stepwise condensation model. We take, as a starting point, a small spherical cluster of "ice-like" liquid water of volume v, undercooled to a temperature T, corresponding to a degree of undercooling $\Delta T = (T_m - T)$, where T_m is the *equilibrium* freezing point of water (*i.e.* the melting point of ice). The free energy of condensation is given by

$$\Delta G_c = RT \ln(p_{ice}/p_{liquid}) \qquad (1)$$

where p_{ice} and p_{liquid} are the sublimation pressure of ice and the saturation vapour pressure of undercooled water, respectively. For any temperature below T_m, ice is the stable phase, so that $\Delta G_c < 0$. Opposing the growth of the cluster are the forces due to surface tension (σ) between the liquid and the cluster surface. For a cluster of radius r, the net free energy accompanying the liquid → solid condensation is

$$\Delta G_{l\rightarrow s} = 4\pi r^2 \sigma - (4/3)\pi r^3 \Delta G_c \qquad (2)$$

The value of r at which $\Delta G_{l\rightarrow s}$ becomes negative, *i.e.* the cluster size for spontaneous crystal growth, is obtained as r^* by differentiation of Equation (2):

$$r^* = -2\sigma/\Delta G_c \quad \text{and} \quad \Delta G^*_{l\rightarrow s} = 16\pi r/3(\Delta G_c)^2 \qquad (3)$$

The nucleation relationships for water at 233 K (−40°C) are shown in Figure 6. The number of critical nuclei in a given volume of water, $n(r^*)$,

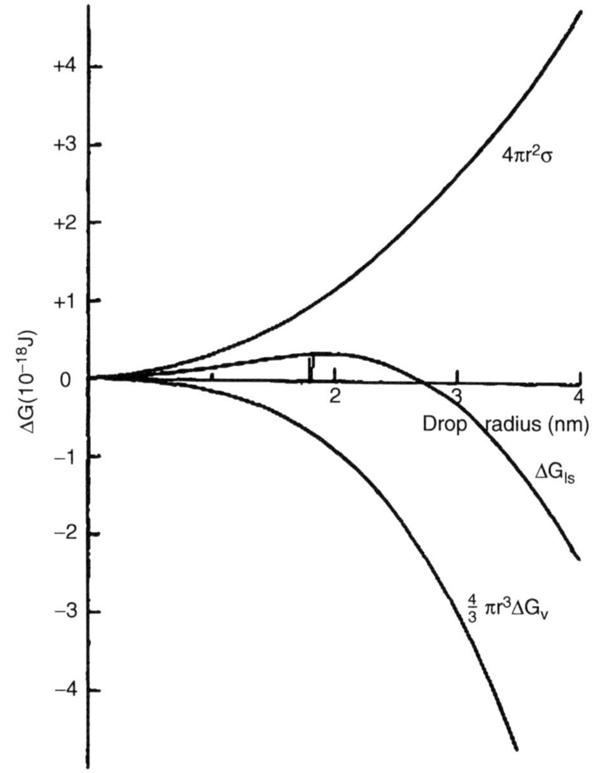

Figure 6 *The volume free energy of homogeneous nucleation of undercooled water at −40°C as a function of the cluster radius at −40°C (see Equation (2)). According to Equation (3), r^* = 1.85 nm. The critical nucleus then contains ca. 200 molecules. The value for σ was obtained from the nucleation data of Michelmore & Franks*[23]

at a given temperature is given by

$$n(r^*) = n_{\text{liquid}} \exp(-\Delta G_c/kT) \tag{4}$$

where n_{liquid} is the number of molecules per unit volume in the liquid phase.

Apart from the dimensions of critical nuclei under varying conditions, it is also important to consider the kinetics of nucleation, especially relative to the kinetics of crystal growth. The nucleation rate (J) is commonly estimated by means of transition state theory. With the aid of the "best" available values for transport properties of undercooled water,[21] it has been estimated that in the neighbourhood of −40°C, J increases rapidly with decreasing temperature, by about a factor 20 per degree; nucleation is thus a well-defined event that is hardly affected by the rate of cooling.[22]

Figure 7 shows the relationship of the mass of liquid water, which is likely to contain at least one active nucleus, as a function of temperature. It is seen that in the range 233–243 K a temperature is reached at which 1 g of water is likely to contain at least one such nucleus, capable of

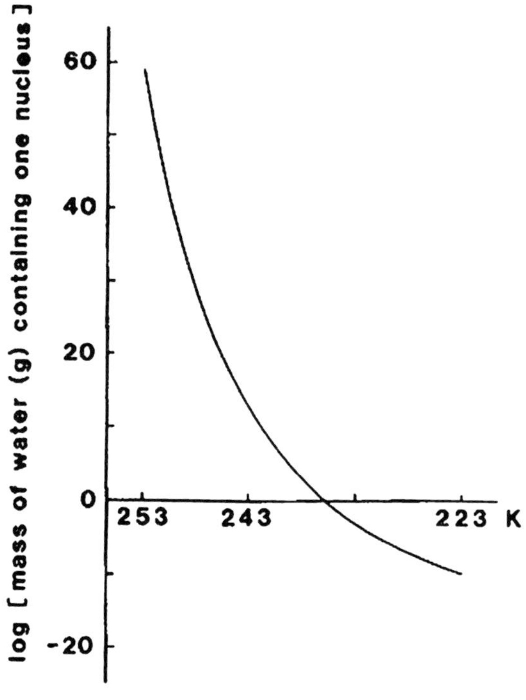

Figure 7 *Mass of water (g) likely to contain one critical nucleus, as a function of temperature, assuming a homogeneous nucleation mechanism. Reproduced from Franks,[17] with permission from Cambridge University Press*

initiating freezing. Although estimates vary considerably, some dimensions and concentrations of such nuclei in water are summarised in Table 1.[17] Thus, if a practical aim is to stabilise undercooled water, then the total mass should be subdivided into small droplets, say of 5–10 nm radius; this would yield *ca.* 10^{18} drops g^{-1}. Referring to the results in Table 1, the proportion of drops susceptible to freezing would then increase from 10^{-450} at 263 K to 10^{-18} at 233 K.[175]

It will now be apparent that the size of an active nucleus decreases with decreasing temperature, thus increasing the probability of their existence at low subzero temperatures. At the same time, the self-diffusion rate also decreases with decreasing temperature, thus increasing the lifetime of the nucleus once it has been formed. Taken together, these results indicate that nucleation rates increase sharply with decreasing temperature in deeply undercooled solutions. In fact, for water droplets of nm radii, there appears to be a limiting temperature, *ca.* −40°C, referred to as the "homogeneous nucleation temperature" (T_{hom}), at which the vast majority of drops will contain at least one active nucleus, capable of initiating ice crystal growth.

Nucleation, although it can be treated as a first-order kinetic process, does not display the typical Arrhenius behaviour, since the rate increases with *decreasing* temperature. The rate J of nucleus generation can be adequately expressed in terms of a reduced temperature and degree of undercooling by

$$J = A\,\exp(B\tau_\theta) \tag{5}$$

where $\tau_\theta = [\theta^3(\Delta\theta)^2]^{-1}$, $\theta = T/T_m$ and $\Delta\theta = (T_m - T)/T_m$.

Nucleation rates, when plotted according to Equation (5), confirm that the relationship can adequately account for the process over several orders of magnitude in J, see Figure 8.[23] The experimental window for nucleation measurements of ice in real time is narrow, because of the extremely high sensitivity of J to temperature and the degree of undercooling. In the temperature range $-35 > T > -45$°C, ln J changes from approximately 20 to −3, with a peak value of 34 at −40°C.[24]

Reference was made earlier to the use of undercooled water in enhancing the long-term stability of enzymes in solution and the viability of isolated cells. The data in Table 1 and Figure 8 confirm that clean, undercooled water at −20°C, dispersed in the form of small droplets, is most unlikely to freeze, and can be safely used as storage medium for many years.

First-order phase transitions can be of several types, *e.g.* vapour → liquid, vapour → solid, liquid → solid, solid → solid. In practice, they are brought about by changes in pressure and/or temperature. Figure 9

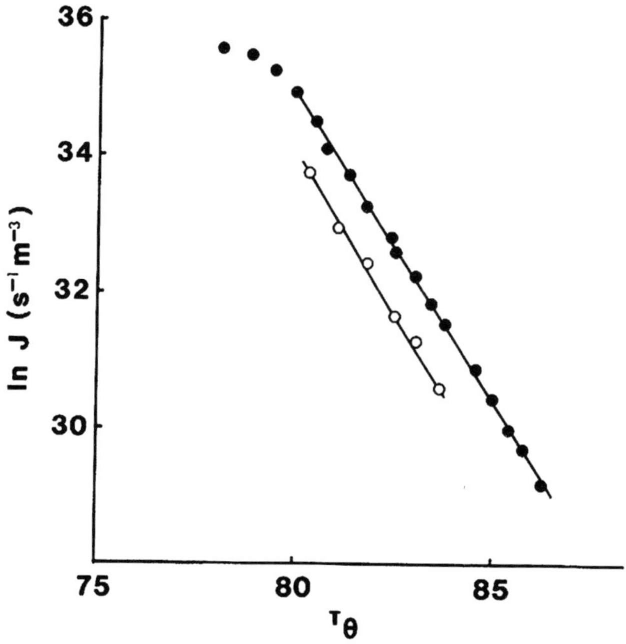

Figure 8 *Temperature dependence of ice nucleation rate in undercooled water, according to Equation (5), from DSC data obtained in the isothermal (○) and scanning (●) modes. Adapted from Michelmore & Franks*[23]

Table 1 *Dimensions and concentrations of active nuclei in undercooled water*[17]

Temperature (K)	*Number of molecules per nucleus*	*Nuclei* (g^{-1})
263	15,900	2.3×10^{-315}
243	566	3.8×10^{-14}
223	122	7.1×10^{9}

shows the partial equilibrium phase diagram of water. The high-pressure polymorphic forms of ice are omitted. They are of no importance to freeze-drying, because they are stable only at pressures in excess of 200 MPa; some of them cannot coexist at all with liquid water. Included in Figure 9 is the pressure dependence of the nucleation temperature. The pronounced effect of pressure on the nucleation rate, as expressed by the nucleation temperature, may help to resolve the question: what are the structural features of an active nucleus? Given the low density of liquid water and ice-Ih, the effect of pressure would be to distort any specific molecular configurations that are required for the promotion of ice growth. Figure 9 demonstrates a so far unexplained parallelism between

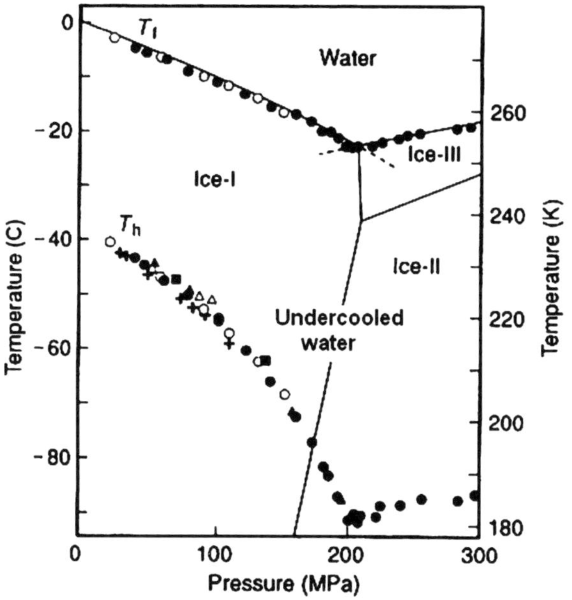

Figure 9 *Partial solid–liquid phase diagram of H_2O, including the corresponding homogeneous nucleation temperature T_h. Reproduced from Franks[17] with permission from Cambridge University Press*

the effects of pressure on the ice melting point T_m and the nucleation temperature T_h, such that at a given pressure, T_h is depressed by twice the amount of the lowering of T_m. It also appears from Figure 9 that the undercooling limit of pure water is –90°C.

Turning briefly to the nucleation of ice in undercooled aqueous solutions, we find that here too, the effect of the molal concentration is exactly analogous to that of the pressure, *i.e.* $\Delta T_h = 2\Delta T_m$. Once again, the reasons are not understood, since the phenomena underlying the two temperature depression effects are quite unrelated. ΔT_m is the well-documented osmotic freezing point depression, whereas ΔT_h is the manifestation of a kinetic rate process. It is also unclear why this process should be identical for equivalent molar concentrations of widely different types of solutes as illustrated in Figure 10.

Reference has also been made to "clean" or "purified" water. In fact, all the numbers shown in Table 1 and Figure 7 are based on the *homogeneous* nucleation model, according to which the generation of nuclei is solely the result of random density fluctuations within the liquid phase. It is, however, well known that nucleation can be catalysed by internal and external factors, in which case the growth and persistence, possibly also the dimensions and structures, of critical clusters are

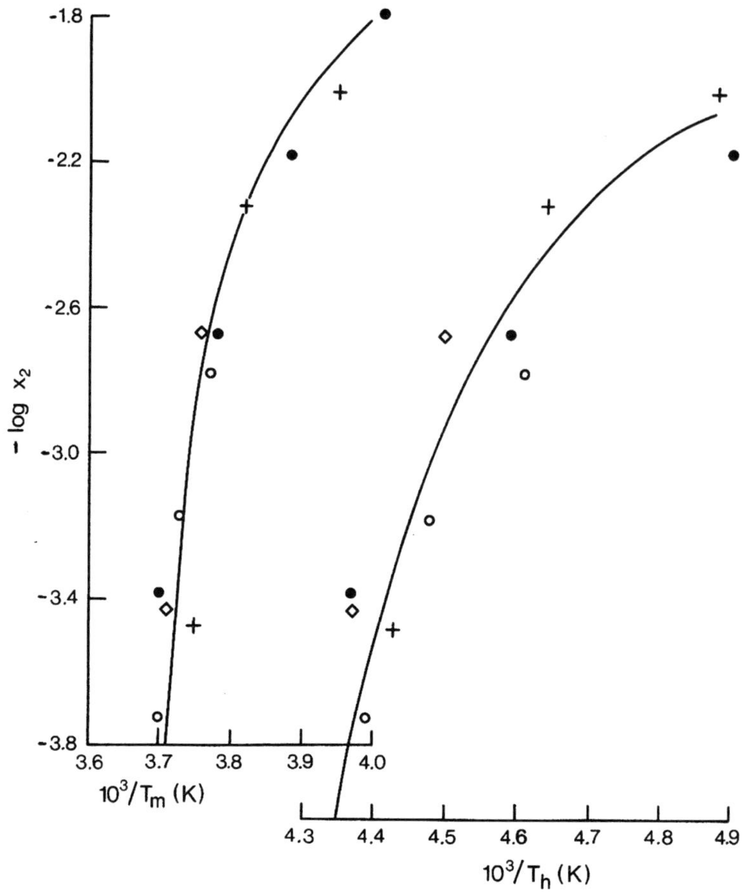

Figure 10 *Depressions of equilibrium melting point (T_m) and homogeneous ice nucleation temperature (T_h) produced by solutes: ∘, glucose; +, NaCl; •, ethane diol and ◇, urea. Reproduced with permission of Kluwer from Franks*[43]

affected. The practical result is that the degree of undercooling that can be achieved is then reduced, and freezing occurs at a higher temperature than would be observed in a rigorously purified system. In practice, millilitre volumes of water cannot be purified to such a degree that homogeneous ice nucleation is achieved. Even sterile-filtered "water for injection" contains "contaminants" that limit the accessible degree of undercooling in bulk to *ca.* −15°C, compared to −40°C for T_{hom}.

Much effort has gone into establishing what particular properties of a substrate are significant in rendering it active as an ice nucleation catalyst.[20] Originally, it was believed that a match of the crystal structure with that of ice was important, a hypothesis that resulted in the choice of silver iodide to promote nucleation of water and ice in supersaturated water vapour. The present consensus, however, identifies three factors that determine the nucleating efficiency of a substrate: (1) it should have a small lattice mismatch with ice, (2) it should have a low surface charge and (3) it should not be wetted by water. The heterogeneous nucleation temperature, T_{het}, is not as easily defined as T_{hom}. It has been suggested that the two processes can be related by

$$\Delta G_{het} = \Delta G_{hom} f(m, R)$$

where R is the radius of the (spherical) catalytic particle and m is a wetting parameter that describes the relative ease of wetting of the particle by ice and undercooled water, respectively, with m varying as $-1 < m < +1$. For maximum catalytic efficiency, $R > 10$ nm and $m \sim 1$.

Great interest was generated by the discovery, some 25 years ago, that enhanced heterogeneous nucleation is one of the mechanisms employed by some freeze-tolerant organisms, in particular insects and plants, to minimise the undercooling of their body fluids when the environmental temperature drops to below T_m. This enables the freezing of the plasma to proceed under quasi-equilibrium conditions where lethal cytoplasmic freezing can be avoided. An early, well-documented example is the alpine plant *Lobelia telekii*. It grows on the slopes of Mount Kenya, at temperatures that fluctuate daily between −10 and +10°C. Its efflorescence contains a potent ice nucleation catalyst that substantially suppresses undercooling. When the air temperature falls to *ca.* −0.5°C, freezing occurs and the latent heat of crystallisation released prevents further falls in the temperature of the plant during the night.[25] The catalyst, probably of polysaccharide origin, has also been shown to be active *in vitro* (see Figure 11), where it is seen to inhibit the undercooling of microdroplets of saline solution (H. Levine, unpublished data).

The environmental implications of biological ice nucleation are still a subject of intensive research. For instance, ice-nucleating ability has been observed in spider web fibroin,[26] but it is uncertain whether this process has any ecological consequences. Unfortunately, no agreement has yet been reached on nomenclature. Care is required to distinguish between terms such as "nucleation temperature", "kinetic freezing point", "supercooling point", and their multifarious interpretations.[27]

Heteronucleation does not necessarily require the presence of foreign solid matter in the bulk or at the surface of the undercooled liquid. It can also be induced by a sudden compression, an electric field or irradiation. The mechanism whereby extraneous factors can influence the nucleating potential within an undercooled liquid is completely unknown. What is important, however, is that at any given temperature, $J_{het} \gg J_{hom}$.

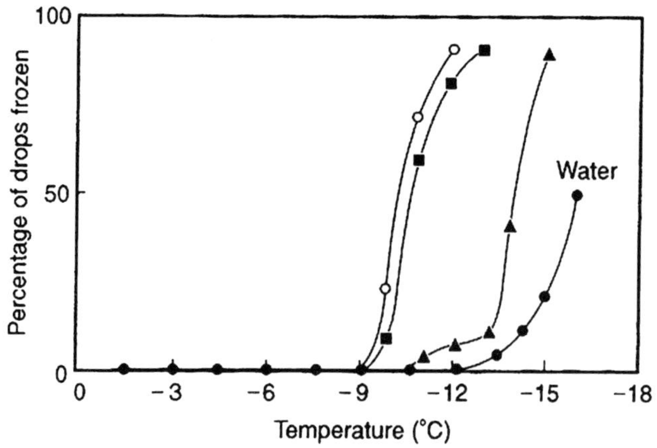

Figure 11 *Undercooling and droplet freezing of water in saline solution (●), aqueous solution of polygalacturonic acid (▲), fluids extracted from the efflorescence of L. telekii (■) and their polysaccharide fraction (○). Reproduced with permission from the Royal Society of Chemistry from Franks*[44]

4.5 Ice: Crystallisation and Crystal Habits

In contrast to *nucleation*, the *growth* of ice crystals in undercooled water is extremely rapid and displays Arrhenius kinetics. The temperature dependence of the overall freezing rate is thus given by a combination of two opposing processes in which the nucleation rate increases with decreasing temperature, whereas ice crystal growth increases with increasing temperature. The freezing process, as it occurs in an aqueous solution, is affected by additional variables: temperature, pressure, solute concentration, catalytically active "impurities", rates of cooling and rewarming, where appropriate, and thermal history (annealing).[28] The crystal habits of ice are extremely complex so that, depending on the environmental conditions, needles, feathers, spherulites or disc-type crystals can be generated.[20,29] On annealing, they all eventually tend to convert to the stable form, the familiar hexagonal stellar dendrites.

A distinction must be drawn between equilibrium freezing, as it occurs under conditions of minimum undercooling, *i.e.* $\Delta T \approx 0$, and perturbed freezing, as observed for rapidly cooled solutions, where $\Delta T \gg 0$. For purposes of industrial-scale freeze-drying, the former process is of greater significance. Cooling rates are invariably low where heat transfer takes place between a liquid in a vial and a cooled metal shelf; typical cooling rates are of the order of 1–30°C min^{-1}. Perturbed freezing can be achieved when small volumes, *e.g.* droplets, are sprayed into a suitable cryogen or against a cold metal plate. In that case, cooling rates well in excess of 1000°C s^{-1} can be reached, *i.e.* $\Delta T \approx 40°$. With ultra-fast quenching methods, thin films or microdroplets of liquid water can even be vitrified.[30]

4.6 Nucleation and Crystallisation of Ice and Solutes in Aqueous Solutions

In the following discussion, emphasis is placed on the classes of solutes that are of importance in pharmaceutical freeze-drying; they include salts and some water-soluble organic compounds, specifically those that can act as hydrogen-bond donors and/or acceptors. As shown in Figure 10, solutes exert a major influence on ice nucleation and, hence, also on ice crystallisation rates and crystal habits. An additional factor to be taken into account is the fate of the solutes during freeze-concentration.

According to the equilibrium phase behaviour of the NaCl–water system, see Figure 12, $NaCl \cdot 2H_2O$ precipitates (crystallises) from the part-frozen solution at −21°C, the eutectic point, T_e; the NaCl concentration in the residual liquid phase, *i.e.* its saturation solubility at this temperature is *ca.* 4 M. T_e is defined formally as the temperature below which no (liquid) solution phase exists in equilibrium with any crystalline phase(s); under equilibrium conditions the system below its eutectic point consists *entirely* of a mixture of ice and $NaCl.2H_2O$ crystals.[†] The dihydrate is not stable, in the sense that it does not melt at a well-defined temperature. Instead it decomposes at a so-called peritectic point (274 K) into anhydrous NaCl and ice. Of even greater importance than the crystallisation of water is the fate of the solutes during freezing, because the ice will eventually be removed by sublimation, whereas the remaining solutes will make up "the freeze-dried product". The crystallisation of the salt from its saturated solution also depends on the existence of a given density of nuclei on which Na^+ and Cl^- ions, as well as water molecules, can condense, thus enabling the $NaCl \cdot 2H_2O$ crystals to grow. Just as in the case of ice, the formation of nuclei depends on random density fluctuations within the body of the now highly concentrated liquid phase. In practice, it is found that few salts will spontaneously crystallise at T_e, so that a certain degree of supersaturation (≡ undercooling) is unavoidable, directly analogous to the transient undercooling of water. Such delayed crystallisation (precipitation) of solutes from a freezing solution, giving rise to supersaturation, is a common phenomenon and has important implications for successful freeze-drying of biopharmaceuticals; these are discussed in more detail in the following chapters.

[†] T_e is very carefully defined here because the commercial freeze-drying literature contains many instances in which a glass temperature is mistakenly referred to as a "eutectic temperature".

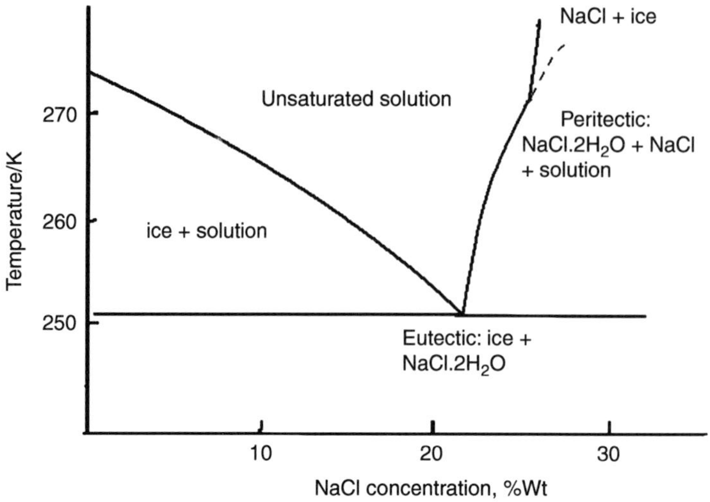

Figure 12 *Solid/liquid phase diagram for the binary water–NaCl system. The dihydrate that separates at the eutectic point is unstable, i.e. it has no melting point but decomposes into anhydrous crystalline NaCl and water (ice) at a peritectic point*

An important feature of transient supersaturation is its duration or, in other words, the rates of *solute* nucleation and crystal growth. Some of the factors that influence these rates have already been discussed for the case of pure water. For crystallisation from solutions, two additional factors are the viscosity of the residual freeze-concentrated liquid phase and probably also the configurational complexity of the crystal structures of the solutes involved. Thus, for NaCl, a degree of supersaturation of up to 6 M, reached at *ca.* −25°C, seems reasonable, at which temperature the salt is likely to crystallise spontaneously. For a typical cooling rate of 10°C min^{-1}, the duration of supersaturation would then be *ca.* 5 min, with the NaCl concentration transient reaching 6 M!

The situation is totally different for many organic molecules, especially polyhydroxy compounds (PHCs), which do not crystallise at all from a freezing solution in real time, even at very high degrees of supersaturation and very low cooling rates. Instead, water continues to freeze beyond the notional eutectic point, and eventually the residual solution vitrifies at some subzero temperature. This behaviour makes them particularly useful as stabilisers of labile biomolecules during freeze-drying and subsequent storage. Vitrification, also known as "amorphisation", can occur inadvertently or it can be planned by expert formulation development. In fact, the choice of excipients is frequently determined by their respective abilities to crystallise or to maintain the dried product in a stable amorphous state. The materials science of water-soluble glasses, which forms an important feature of pharmaceutical freeze-drying, is explored in detail in several later chapters.

An analysis of crystallisation rates is conveniently performed in terms of the so-called time–temperature–transformation (TTT) curves, which relate the time taken to crystallise a given fraction of the undercooled liquid or the supersaturated solution to the temperature. Experimentally, the crystallisation rates are measured by quenching the liquid phase to some predetermined temperature T and measuring the time taken for the solid to crystallise at that temperature, either by monitoring the latent heat of crystallisation or by microscopic observation. The volume fraction $\phi(T)$ that crystallises out in time t is given by one form of the Avrami equation:

$$\phi(T) = (\pi J u^3 t^4)/3 \tag{6}$$

where J is the rate of nucleation and u the rate of crystal growth. Examples of TTT curves for glycerol from its melt and of ice from undercooled water are shown in Figure 13. The so-called nose of the TTT curve indicates the maximum crystallisation rate that can be attained. Any cooling rate high enough not to intersect the TTT curve will, therefore, lead directly to vitrification. In the case of water, the maximum rate of ice growth occurs near $-13°C$ at *ca.* <1 ns nm^{-1}, indicating that cooling rates in excess of 5000°/s need to be applied in order to prevent freezing. Cooling rates ($-$ dT/dt) are proportional to $(\kappa/C_p d)$, where κ is the thermal conductivity, C_p the heat capacity and d the density. Both water and ice have low thermal conductivities and large heat capacities. It therefore becomes difficult to achieve high cooling rates, especially for bulk samples, but vitrification has been achieved with specialised quenching techniques applied to microdroplets and thin films.[31, 165]

Inspection of simple binary solid–liquid phase diagrams, as shown in Figure 1, indicates that, if a solution, the composition of which lies to the right-hand side of the eutectic point, is cooled, the salt, and not water, will be the component that crystallises first. A TTT curve for salt precipitation could thus be constructed, but the rates would be much lower than those shown in Figure 13 for ice crystallisation. The importance of salt crystallisation during freeze-drying is further elaborated in Chapter 5.

Yet two further aspects associated with ice growth in a freezing solution deserve mention. For the sake of simplicity, we consider the advance of a planar ice front during controlled cooling, shown in Figure 14 for a 2% w/w solution of $NaMnO_4$. A thin film is held between two thin glass plates set 10 μm apart and is seeded with an ice crystal and cooled inwards from

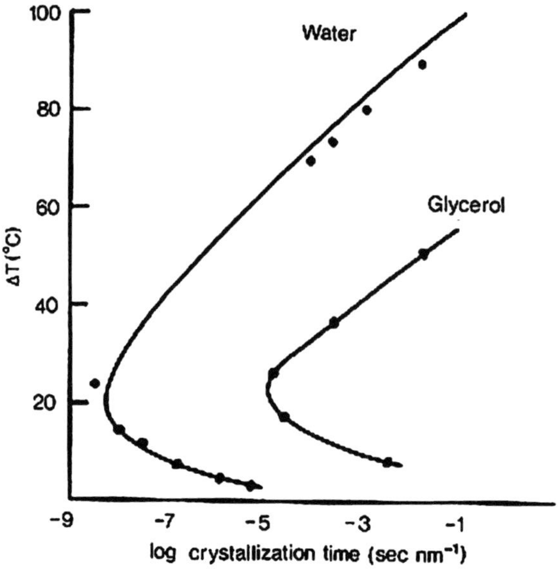

Figure 13 *TTT curves for the crystallisation of water and glycerol. For details, see text*

the two edges‡.[28] Initially, a planar ice front is maintained. As the front moves into the body of solution, a concentration gradient is established, because diffusion of the salt away from the freezing interface cannot keep up with the rate of advance of the ice front. The concentrated solute produces a freezing point depression, over and above that due to the *equilibrium* colligative effect. This effect is termed "concentration polarisation".[168] At some stage (between 72 and 84 s in Figure 14), the ice/solution interface begins to show the first signs of instability. Freezing is then no longer unidimensional, and dendrites begin to form, eventually producing the familiar dendritic ice structure. Finally, the residual supersaturated solution phase is forced into spaces between the dendrites, where either eutectic crystallisation or vitrification will occur. It is of importance to realise that, even under rigorously controlled cooling conditions, *i.e.* thin films or microdroplets cooled at low rates, the process becomes uncontrollable after the appearance of interfacial instabilities, due to solute concentration polarisation ahead of the solid/liquid interface. It follows that under conditions of bulk freezing of *unseeded* solutions (*i.e.* undercooling and random nucleation), the freezing process cannot be controlled.

‡ $NaMnO_4$ was chosen because (1) its eutectic temperature lies close to that of NaCl and (2) the concentration can be monitored spectrophotometrically in the visible range.

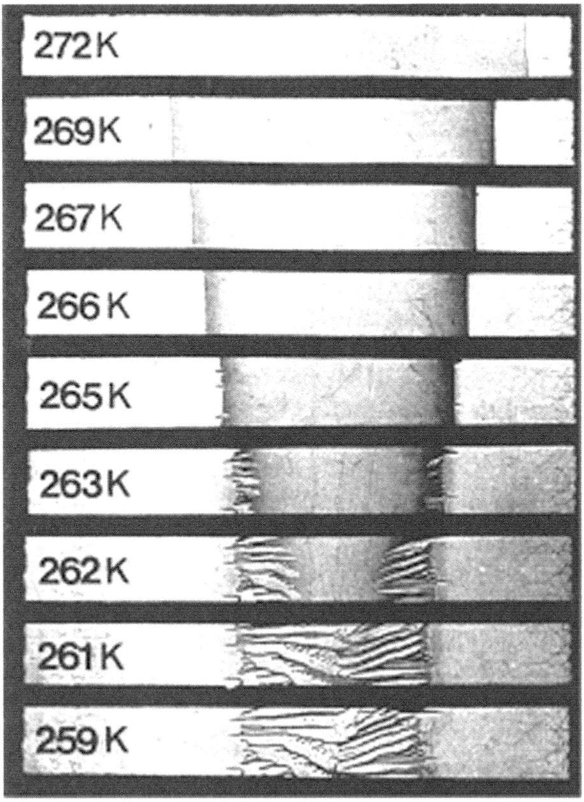

Figure 14 *Unidimensional freezing and its eventual perturbation in a $NaMnO_4$ solution, illustrating the effect of concentration polarisation ahead of the advancing ice front. Reproduced from Körber*[28]

The other aspect refers to structural changes that may occur within the ice matrix during rewarming, even well below T_m. The process, referred to as "maturation" or "annealing", describes a disproportionation whereby large crystals grow at the expense of small crystals, producing a coarsening of the frozen matrix, *i.e.* a reduction in its porosity. Maturation is of particular concern to frozen food producers, because the threshold for the ability of the human tongue to discriminate between individual ice crystals is *ca.* 50 μm. In ice cream, this gives rise to a "grainy" texture that can, however, hardly be prevented when a carton is removed from and replaced into the freezer several times during its period of use. An example of ice maturation in frozen meat is shown in Figure 15.

Ice maturation becomes of relevance to freeze-drying, where a preparation may have been optimally frozen but is then left in the freeze-drier at some suboptimal temperature before the drying cycle is commenced. Figure 16 shows the appearance of a frozen polyvinyl pyrrolidone (PVP)

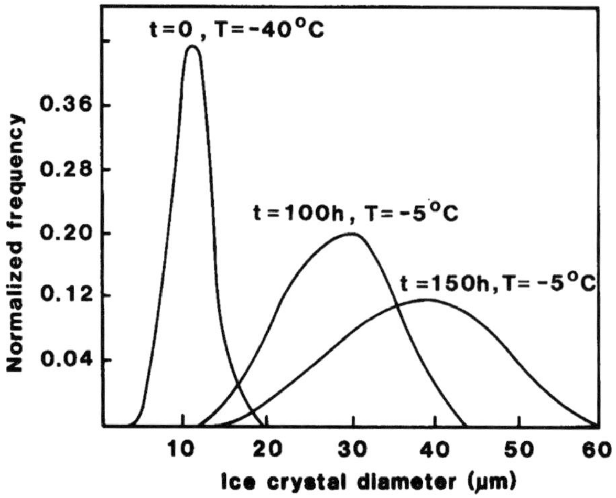

Figure 15 *Effects of temperature and frozen storage on ice crystal dimensions in details. After Calvelo.*[45] *Reproduced from Franks*[17] *with permission from Cambridge University Press*

solution, quench cooled and kept at −196°C.[32] The average ice crystal diameter is 100 nm. In the particular study shown, after allowing the temperature to rise to −20°C and repeated quenching, a process that took <1 min, the mean ice crystal diameter had already risen to >10 μm.

In freeze-drying technology, maturation of the frozen matrix has certain advantages, *e.g.* a reduction of the required drying time, but these must be weighed against distinct disadvantages, such as a decreased rate of reconstitution of the dried cake at the point of use. Note also the curious spherical appearance of the freeze-concentrated polymer solution phase. Such supramolecular spherical assemblies have also been observed with other polymers, *e.g.* polyvinyl alcohol and polyethylene glycol (PEG) in frozen solutions. It is not obvious how an ostensibly homogeneous solution is converted into such uniform hydrated spheres of submicron dimensions during freeze concentration, but the possibility of a demixing process (lower critical solution temperature, LCST) in such systems cannot be ruled out.

4.7 Polymorphic Transitions, Hydrates and Transient Hydrates

The basic outlines of the solid/liquid phase diagram of water have gradually been developed, beginning with the advent of X-ray crystallography in the 1930s and with a further boost from more recent advances

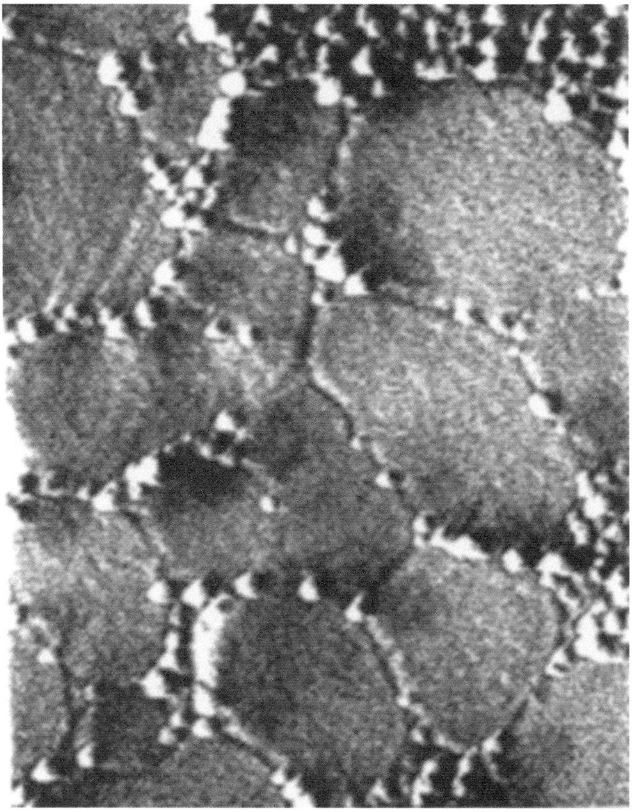

Figure 16 *Transmission electron micrograph of a quench-frozen and etched 20% solution of PVP. Grey areas correspond to prior positions ice crystals; their dimensions are ≈ 100 nm. The solution phase (PVP + unfrozen water) is confined to the interstices between ice crystal; see bottom slide in Figure 14. Reproduced with changes from Franks et al.*[32]

in neutron diffraction.[33] Although an increasing number of ice polymorphs that can exist under various pressure/temperature regimes have been identified and their crystal structures established, only "ordinary" or hexagonal ice (ice-Ih) is of significance in the freeze-drying process. This is the familiar form that is obtained when water in bulk is slowly cooled to subzero temperatures. A comparison of the physical properties of cold liquid water with those of ice-Ih demonstrates some of the anomalies of H_2O. Thus, freezing takes place with a volume expansion, combined with a 50% reduction in the molar heat capacity. Both these changes have implications in freeze-drying technology.

Whereas in the preceding sections of this chapter the emphasis has been on ice nucleation and crystal growth, an aqueous solution, destined for freeze-drying, will also contain other chemical species that might be able to crystallise during the cooling process, or even during subsequent annealing and/or rewarming. As in the case of ice, their crystallisation will depend on their respective nucleation kinetics. For multicomponent systems, a general rule is that the crystallisation of any one component is likely to be "poisoned" (*i.e.* retarded) by other components. This topic will be treated in more detail in Chapter 6 because it is of importance in pharmaceutical formulation development.

As regards pharmaceutical product development, conventional wisdom has it that crystallisation should be the desired end result of freeze-drying. The rationale is that the crystal is the state of minimum energy, and it is therefore the most stable state. This is, however, a gross oversimplification. Many drug substances, and even more excipients, can exist in a number of polymorphic forms and/or as hydrates. The actual species that is produced during drying is likely to be influenced by the particular processing conditions and may then not be the most stable polymorph.[34] It is also possible that crystalline states appear in the form of hydrates or mixtures of different polymorphs. Even more important is the possibility that, where a chemically complex solid product is shown to be crystalline, solid–solid transitions might occur during storage, probably giving rise to a more stable product, but also one that might have other undesirable properties, such as low solubility or enhanced chemical reactivity.

The behaviour of the lower alcohols in aqueous solution merits a special comment. Despite their apparent complete miscibility with one another and with water, the alcohols display an extremely complex series of eutectic, peritectic and hydration phenomena, which have usually been hidden from earlier observers because they appear at low temperatures. As one example, consider the mixture water–*tert*-butanol shown in Figure 17.[35] A stable hydrate of composition $A \cdot 2H_2O$ and $T_m = 0.55°C$ is clearly shown, and there are also indications of several peritectic transitions and metastable hydrates. Similar complexities have been described for aqueous solutions of methanol, ethanol and propanol.[36]

4.8 Vitrification as Avoidance of Crystallisation

Apart from the ice polymorphs, other types of solid but non-crystalline water exist, and they apparently form the most abundant state of H_2O in the universe, where they are found adsorbed onto interstellar dust particles that eventually form comet tails. Amorphous, glassy water is mentioned here only for the sake of completeness, because its properties

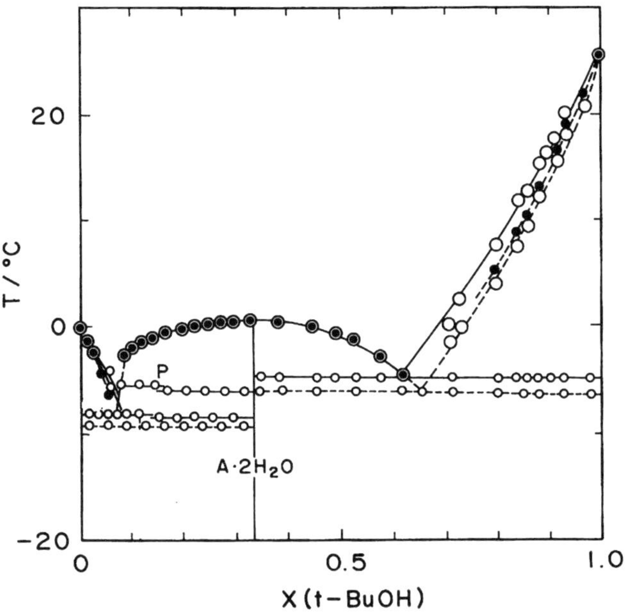

Figure 17 *Solid–liquid phase diagram of the system water-tert-butanol; X = mol fraction concentration and P denotes a probable peritectic transition. A.2H_2O is a stable crystalline hydrate, with T_m = 0.55°C. The two eutectic points correspond to −8.52°C (x = 0.065) and −4.55°C (x = 0.62). Adapted from Takaizumi*[35]

are currently still subjects of intense research activity[37] and dispute.[169] Two "states" of amorphous water have been identified; they differ in density and are formed by the rapid quenching of liquid water, by the deposition of water vapour on cold surfaces, or by the compression of hexagonal ice at low temperatures.

According to some researchers, the two states are in fact two distinct phases, with real coexistence boundaries.[38] If this claim can ever be verified, then it follows that there should exist a second critical point for water, akin to the critical point at 374°C, where the liquid and gas phases coalesce. It may, however, be impossible to confirm the existence of such a point by experiment, because it would lie well below T_{hom}, where freezing cannot be avoided.

While amorphous water as such may not figure in freeze-drying technology, the opposite is true for many of the other components that make up the finished pharmaceutical preparation. In fact, the realisation of the importance of amorphism in pharmaceutical product development has led to the birth of a new scientific discipline: pharmaceutical materials science. Successful freeze-drying is firmly based on the same principles as classical materials science, *e.g.* silicates, borates, metal alloys, *etc.*, but applied to water-soluble or water-sensitive materials.[39] Thus, as will be made clear in the following chapters, the *avoidance,* rather than the promotion of crystallisation in pharmaceutical processes, may be a worthwhile objective. It has, for instance, been reported that insulin in the freeze-dried amorphous solid state is substantially more stable than in the crystalline form,[40] which led the authors to doubt whether the "rule" that the crystalline phase always provides the higher stability towards degradation is universally applicable. The following chapters explore the relative merits of the two most important physical states in freeze-drying technology.

4.9 Freeze-Drying of Complex Biological Materials

Although not strictly germane to the subject matter of this book, mention should be made of attempts that describe the application of freeze-drying to the long-term storage and eventual functional recovery of a range of biological materials of much higher complexities than isolated biomacromolecules or formulated solutions that contain these molecules. Such materials can be of plant or animal origin and are, in increasing order of complexity: single cells, cell clusters (*e.g.* embryos), tissues, organs and even simple organisms. (Excluded are biological materials, such as dried flowers, where there can be no question of "recovering functional activity".) Although there are journals in which freeze-drying reports appear and there are even patents that protect such inventions, to the author's knowledge, such claimed "inventions" have not yet been successfully reduced to practice, clinical or otherwise.

However, low-temperature preservation by deep chill and dormancy is indeed a very effective method to preserve *in vivo* functioning. The technical problems are then to find methods of preventing water from freezing, or at least to control the degree of freezing. The principles and techniques have been described earlier in this chapter. Results are shown in Figures 2 (*in vivo*) and 3 (*in vitro*). To extrapolate such results to achieve long-term laboratory preservation of "real" biological material in the unfrozen state, the same principle, *i.e.* inhibition of ice nucleation, must be employed. Freeze-drying cannot therefore be the strategy of choice, but freeze avoidance is. Success has been achieved with cells and even some plant tissues, by encapsulating the material in aqueous microdroplets and emulsifying the aqueous phase in an oil, the molecules of which cannot act as catalysts for ice nucleation.[16,41] Figure 18 shows two such emulsions of human erythrocytes (left) and *Saccharomyces cerevisiae* cells (right). Both these preparations were stored at −20°C for several weeks, after which

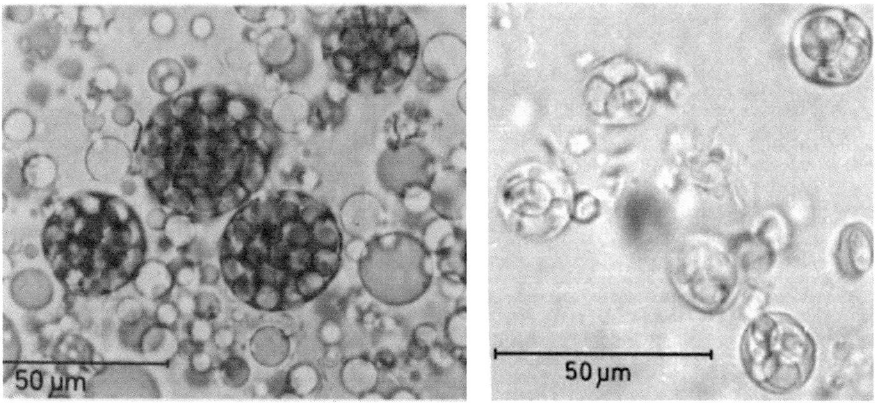

Figure 18 *Photomicrographs of emulsified cells in undercooled water: left, human erythrocytes and right, yeast (S. cerevisiae). The mean diameter of empty water droplets is 7 μm*

they were examined for survival by signs of haemolysis and growth of bacterial activity, respectively.[166] Neither preparation showed signs of deterioration. Similar success was achieved (Mathias & Franks, unpublished results) with tobacco embryos, grown in tissue culture, and stored at −20°C. They were successfully grown into healthy plants. In the preparations shown in Figure 18, analysis by differential scanning calorimetry (DSC) during cooling of the emulsions indicated two freezing events: one of the intracellular water and the second one at a lower temperature, close to −40°C, *i.e.* at T_h. The ratios of the signal amplitudes corresponded exactly to the concentration ratios of intracellular water to empty water droplets.

The undercooling technique can be considerably refined and is thus a useful means of storing live material or clinical fluid specimens for prolonged periods. The technique is not useful or commercially attractive for the long-term storage and transport of large quantities of material, mainly because of the stringent temperature limits imposed, either by inadvertent freezing below the nucleation temperature of water in the material on the one hand, and an upper temperature, usually < 10°C, on the other hand, at which the emulsion will become unstable and break.

CHAPTER 5

Essential Chemistry and Biochemistry Associated with Low Temperature and Freezing

5.1 Physicochemical Properties of Water-Based Systems Under Conditions of Deep Chill

The influence of temperature on physical properties of water has been discussed in the previous chapter, where it was pointed out that the degree of ionisation (pK_W), in particular, is markedly affected by low temperature. In principle, such changes should be, but seldom are, a factor to be taken into account in the selection of suitable buffers for the freeze-drying of pH-sensitive formulations. Other property changes, directly affected in a major way by chill, and of relevance to freeze-drying, include solubility, viscosity and chemical reaction rates and mechanisms, the latter being of particular importance where coupled reaction cascades may be involved. The physical properties of members of a series of homogeneous aqueous solutions can respond quite differently to temperature changes. Some changes are monotonic (*e.g.* refractive index, dielectric permittivity), while others exhibit extrema (*e.g.* density, compressibility). Some changes are very marked (*e.g.* pH), while others are minimal (*e.g.* hydrogen bond length). The physical properties of water at subzero temperatures were first discussed in detail by Angell,[21] and their biochemical and biotechnological implications have been reported by the author.[17] Solubility and miscibility present particularly interesting cases: the solubility of most substances increases with rising temperature, a behaviour that would be expected, since temperature is a measure of kinetic energy. While this is indeed an underlying universal behaviour of *all* substances, certain classes of compounds do, in addition, exhibit the phenomenon of lower critical demixing, *i.e.* their miscibility with water decreases with *increasing* temperature, giving rise to the well-known closed-loop liquid–liquid phase diagrams, *e.g.* water–polyethylene glycol.[46] Evidence for such phase separations, as they develop during freeze-concentration, has been cited,[47] but the phenomenon remains to be definitely established. If such phase separation can indeed play a role during freezing, some current views on formulation development would need to be significantly modified.

5.2 Cold Inactivation of Proteins

The diverse phenomena surrounding protein stability in solution have been studied in great detail but are still subject to several unresolved problems. In the context of this monograph, emphasis is placed on the cold inactivation of proteins, a process that is quite unrelated to freezing. The possibility of its existence was first suggested by Brandts,[48] based on detailed studies of the combined effects of pH, urea concentration and *high* temperature on the reversible denaturation of chymotrypsinogen, as shown in Figure 1.

The parabolic nature of the thermal stability profile, which for obvious reasons terminates at the freezing point of water, led Brandts to

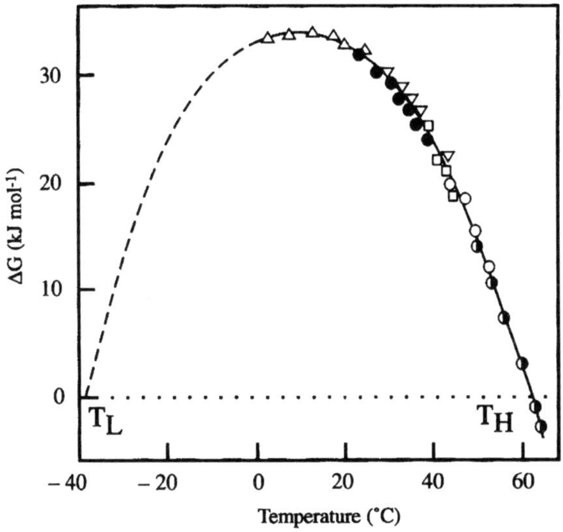

Figure 1 *The effect of pH on the thermal stability of chymotrypsinogen according to Brandts.*[48] *Circular symbols refer to a range of pH values; triangles refer to 2.3 M urea.* T_H *is the well-characterised thermal denaturation temperature.* T_L *is the experimentally determined denaturation temperature in undercooled water Franks*[48]

speculate that in the absence of freezing, a cold denaturation temperature might be expected at some subzero temperature. At the time, the technique of maintaining water unfrozen at subzero temperatures had not yet been developed, and it was not until 1985 that cold denaturation could first be demonstrated experimentally[41] and Brandts' perceptive speculation was put on a firm basis. Since then, the phenomenon of protein cold inactivation has been studied in more detail.[49] For reasons of experimental convenience, all such studies had previously been performed by perturbing a protein simultaneously with temperature in combination with another destabilising treatment, *e.g.* pH, urea, guanidinium chloride, cryosolvents or site-directed mutants. In this manner, cold denaturation could be shifted to temperatures above the freezing point of the aqueous solvent medium. It was, however, later demonstrated that such combined destabilisation treatments often produce ambiguous results. Preferably temperature alone should be employed to study a temperature-related phenomenon.[15] Experimentally, this necessitates the use of undercooled water as a reaction medium, a technique the application of which requires a high degree of technical skill but produces a wealth of information.[49,50]

Perhaps the most important results obtained from undercooling studies on proteins can be summarised as follows:[49]

- Cold denaturation (unfolding, dissociation, inactivation) is 100% reversible, even at high protein concentrations.
- Renaturation is generally slow, but complete and correct.
- The mechanism of cold inactivation in undercooled water differs from that brought about by cryosolvents.
- Cold inactivation is brought about by a weakening of the hydrophobic stabilising effects and a concomitant strengthening of direct hydration interactions between water and polar or ionic amino acid residues, thus upsetting the balance between stabilising and destabilising interactions that are responsible for the commonly observed marginal stability of proteins in solution.
- In its cold-inactivated state, a protein is not markedly susceptible to pH changes or salt effects and is not subject to aggregation.

The actual mechanistic and molecular origin of cold inactivation is of great fundamental interest[51] and also has implications for improvements in targeted protein design. The phenomenon is probably not immediately relevant to the behaviour of frozen solutions. It is also not yet clear whether, in the context of freeze-drying, cold denaturation might have a stabilising role during the initial undercooling of a protein solution.

If this were indeed the case, it would be beneficial in countering the stresses due to freeze-concentration (see later).

5.3 Changing Phase Relationships during Freezing

The fact that freezing is itself a drying process is not universally realised. That may in part be due to the conventional terminology used to describe phase equilibria, and also to the conventional representation of solid–liquid phase diagrams, where the composition of a mixture is plotted as a function of temperature, as shown in Figure 1a in Chapter 4 for the simple binary system water–KCl. The left-hand side of the phase diagram is described as "liquidus", whereas the right-hand side is termed "solidus". Thus, the liquidus measures the equilibrium freezing point depression caused by the addition of KCl to water and the coexistence of ice and the unsaturated aqueous solution; it could equally well be regarded as a measure of the precipitation of water from the saturated solution. Similarly, the solidus could be regarded as a measure (hypothetically) of the freezing point depression of KCl due to added water and the increasing solubility of water with decreasing temperature. The liquidus terminates at the so-called *eutectic point* T_e, which, by the law of the phase rule, is the only temperature at which crystalline water and crystalline KCl can coexist in equilibrium with the (saturated) liquid solution. The phase diagram of the NaCl–water system is of more immediate relevance to freeze-drying because buffered saline solution is commonly used as an isotonic reconstitution medium. On the phase diagram, shown in Figure 12 in Chapter 4, the eutectic point occurs at −21.13°C, at a NaCl concentration of 4 M (23.3% w/w). However, the substance that actually crystallises at T_e is not NaCl, but the dihydrate $NaCl.2H_2O$. That makes the construction of the solidus complicated, because the dihydrate is converted to anhydrous NaCl and water at 1°C at the *peritectic*, because $NaCl.2H_2O$ decomposes below its hypothetical melting point. Because of its universal importance, and because it is inexpensive and easily purified, the NaCl–water system has been studied over a wide range of pressures and temperatures.[52] Additional eutectic points have been detected, but they are of no relevance to processes that occur under ambient or near-ambient conditions. The fact that only one hydrate has been found cannot be taken to mean that other hydrates might not exist. As will be shown later, the presence of previously unknown hydrates can cause severe problems in freeze-drying operations when they inadvertently decompose with the liberation of water. The properties of eutectic mixtures are further explored in Chapter 11.

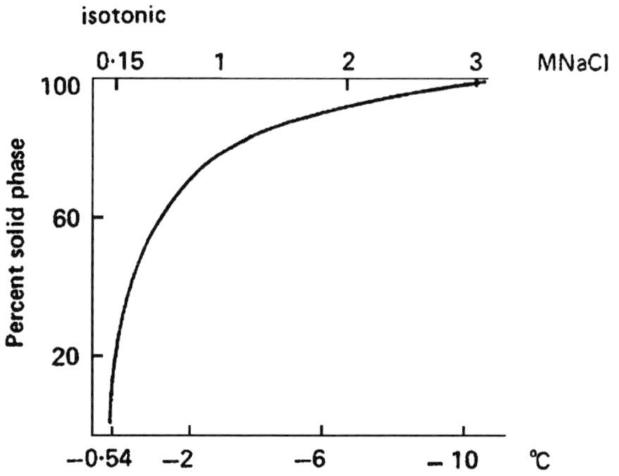

Figure 2 *Unconventional representation of the NaCl–water solid–liquid phase diagram in terms of solution concentration as a function of temperature, beginning at the equilibrium freezing point of an isotonic solution. At the eutectic temperature (−21°C), the saturated solution has a concentration of 4 M, corresponding to a 27-fold concentration increase during freezing*

Solid–liquid phase diagrams can be constructed fairly easily for systems where equilibrium conditions are rapidly established and where assays are simple to perform. The situation becomes complicated when the kinetics of dissolution lie outside the time frame of the experiment. This is unfortunately the case for most substances of relevance to freeze-drying. The problem is further discussed in Section 5.7.

A more dramatic representation of the concentration effect of freezing is obtained by redrawing the NaCl–water phase diagram in a less conventional form, as shown in Figure 2, where temperature (decreasing) and solution concentration (increasing) are shown as functions of the amount of ice formed. Thus, an isotonic saline solution (0.15 M, 0.9% w/w) has an equilibrium freezing point of –0.54°C. Upon cooling the solution to –10°C, most of the water has frozen, and the residual solution now has a concentration of 3 M that, on further cooling to the eutectic point, reaches a limiting concentration of 4 M, *i.e.* a 27-fold concentration increase from isotonic conditions. The process is referred to as "freeze-concentration", but it can equally well be regarded as the (freeze)-drying of sodium chloride from an aqueous solution, because the drying has already been accomplished, even without the removal of H_2O (as ice) from the system. To complete the process, all that remains to be done is the physical removal of crystallised water (ice), and that can hardly be referred to as a drying process.

A corollary of this representation of freeze-concentration is that the process, commonly referred to as "freezing damage" or "freezing

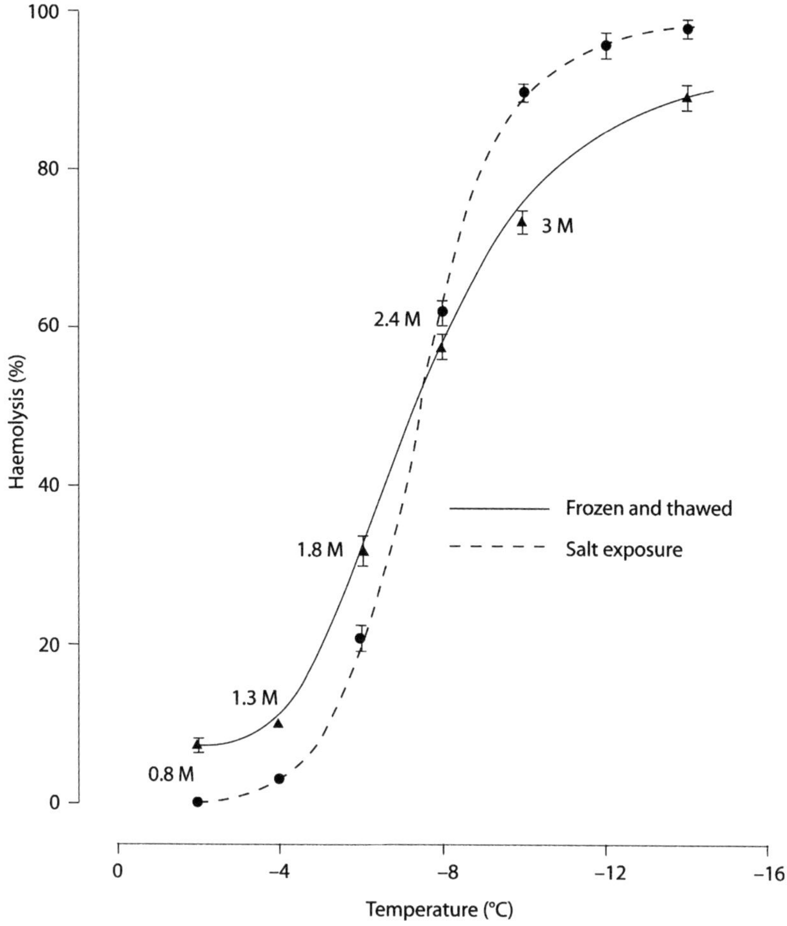

Figure 3 *A comparison of the effects of freezing and salt exposure at room temperature on the haemolysis of erythrocytes. Salt concentrations corresponding to decreasing temperatures are indicated; constructed from data by Lovelock*[53]

injury",[53] and often attributed to ice crystals, is in fact a *concentration* damage that can equally well be induced at ambient temperature and concentrations equivalent to those produced by freezing the solution to the equivalent temperature. The comparison in Figure 3 graphically shows that, when a suspension of human red blood cells in isotonic saline solution is frozen to −8°C, the solution has been freeze-concentrated to 2.5 M. Upon thawing, a 50% haemolysis is observed. The same result is obtained by incubating the cells in 2.5 M of NaCl at ambient temperature.[54] The damage to the cells should therefore be referred to more correctly as "concentration injury". It must, however, be realised that *real* ice crystal damage can become a disruptive and destructive factor during the freezing of vascular tissues.

5.4 Eutectic Crystallisation of pH Buffer Components

Eutectic formation, as predicted by the phase rule, can have a dramatic effect on pH buffer solutions during freezing. In a ternary solution of two salts and water, four eutectic points may, in principle, be identified, of which one is a ternary eutectic (at the lowest temperature), while the other three are binary eutectics of the three pairs of components. Some eutectic data for sodium and potassium phosphate buffer salts with ice are summarised in Table 1.[55,56] For mixtures of the sodium salts, the ternary eutectic temperature is –9.9°C and the mol ratio NaH_2PO_4/ Na_2HPO_4 is 57. In a pH 7 buffer mixture the ratio is, however, 0.72. Thus, in any mixture in which the two salts are not present in their eutectic ratio, the dibasic sodium salt will precipitate first, at –0.5°C, giving rise to a major pH shift. The situation is more favourable for mixtures of the two potassium salts because the ternary eutectic ratio (0.48) lies much closer to the pH 7 buffer composition. The lower eutectic temperature of the potassium phosphate system accounts for the observation that, under normal freezing conditions employed in biochemical studies or food processing, potassium buffers provide a better pH stability than the corresponding sodium phosphate buffer mixtures. Two additional buffer solutions can be prepared by mixtures of Na and K salts; their eutectic data are not included in Table 1.

The situation becomes more complicated when salts can crystallise as hydrates. According to the tenets of the phase rule, additional eutectic points will have to be taken into account for each additional phase. This statement requires modification, where one or several phases are metastable, *i.e.* they decompose before a true melting point is reached. Where such behaviour is observed, the ternary coexistence point between the stable, the metastable and the ice phase with the aqueous solution is referred to as a "peritectic" point. Such behaviour is shown in Figure 13 in Chapter 4 for the NaCl–water system. As pointed out earlier, at the eutectic temperature the salt phase that precipitates is $NaCl.2H_2O$. This dihydrate has a very limited stability; it decomposes (not melts) at a peritectic point of 1°C into the stable anhydrate and water (ice). The impact of unknown or unexpected metastable hydrates on freeze-drying can be significant, because after the completion of freezing, a substantial amount of water may remain "locked up" in a crystalline hydrate of one or several solution components. Depending on its eutectic and/or

Table 1 *Eutectic data of aqueous phosphate solutions*

	T_e (°C)	$C_e(1)$ M	$C_e(2)$ M	Crystal type	$C_e(1)/C_e(2)$
NaH_2PO_4	−9.7	3.42	–	NaH_2PO_4.2W	–
Na_2HPO_4	−0.5	–	0.11	Na_2HPO_4	–
NaH_2PO_4–Na_2HPO_4	−9.9	3.42	0.6	NaH_2PO_4.2W Na_2HPO_4.12W	57 (0.72)[a]
KH_2PO_4	−2.7	0.92	–	KH_2PO_4	–
K_2HPO_4	−13.7	–	2.85	K_2HPO_4.6W	–
KH_2PO_4–K_2HPO_4	−16.7	1.30	2.70	KH_2PO_4 K_2HPO_4.6W	0.48 (0.72)[a]

W, water; (1), mono Na or K salt and (2), di Na or K salt.
[a] Corresponds to pH 7 buffer ratio.
Data from Van den Berg and Rose[55] and Murase and Franks.[56]

peritectic point, water of crystallisation might not be completely removed during primary drying, thus leading to stability problems during storage of the "dried", but not dry product.

An example of a previously unknown hydrate was recently reported for mannitol, a commonly used excipient in pharmaceutical freeze-drying. It was formed inadvertently during the particular drying protocol used.[57] It was identified as a hydrate by X-ray diffraction, differential scanning calorimetry (DSC) and thermogravimetry, and it will be further discussed in a later chapter.

In Chapter 4, reference was made to the general phenomenon of undercooling and the process of nucleation, particularly as it affects the crystallisation of ice. The discussion will now be broadened because *any* solid phase that separates from a liquid mixture is subject to the same physical laws that apply to water. The phase rule, which predicts the *equilibrium* phase coexistence in solid and liquid mixtures, does not take into account the rates at which solid phases can grow within the liquid solution. These rates are governed by the theory of nucleation as discussed in the previous chapter. Since nucleation is a stochastic process and, in practice, can also be affected by cosolutes or even by trace contaminants in a solution, it is impossible to ascribe definite nucleation or crystallisation temperatures to particular substances. As a general rule, however, spontaneous crystallisation *at* the eutectic temperature is rare; hence, that some degree of transient supersaturation is common. An example of such a process, in the form of a simulation, is illustrated for an isotonic saline solution in Figure 4 (Pikal, personal communication). In this example, a cooling rate of 1.25°min^{-1} is applied, and freezing is assumed to occur at −15°C. On continued cooling, as the liquid water content decreases, the NaCl concentration correspondingly increases. No discontinuity (crystallisation/precipitation) is observed at

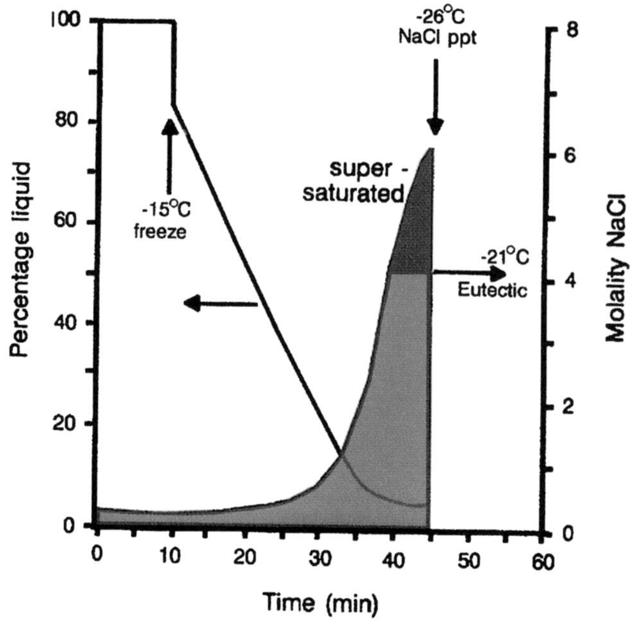

Figure 4 *Simulation of typical concentration changes during the cooling of a NaCl solution showing undercooling of water, ice nucleation and growth, transient supersaturation and final NaCl crystallisation, with the re-establishment of equilibrium conditions*

the eutectic temperature of –21°C. With further cooling, the solution becomes supersaturated in NaCl until, after 45 min, when the NaCl concentration has reached 6 M, nucleation of NaCl becomes effective, salt spontaneously and rapidly precipitates, and eutectic conditions are re-established. Any labile biochemical product, present in this particular solution, would therefore have been exposed to a highly supersaturated salt solution for 5 min.

The actual crystallisation behaviour of any given solute during freezing depends on its nucleation rate, which, in turn, is affected by the following factors:

- Solution volume and concentration
- Cooling rate
- Crystal structural complexity
- Competing solutes (excipients or contaminants).

Under the conditions shown in Figure 4, a supersaturation degree reaching approximately 30% is maintained over 5 min before eutectic equilibrium is established. This indicates that the crystal structure of

$NaCl.2H_2O$ is of a fairly simple nature and that the ionic interactions favouring nucleation and crystal growth are strong compared to ionic hydration interactions.

The situation can, however, be quite different. Consider the system sucrose–water: during freezing, even at moderate cooling rates, sucrose does not crystallise spontaneously at its eutectic temperature, –9.9°C, or thereafter, in real time. With continued cooling, the solution viscosity increases, eventually in a steep, highly non-Arrhenius manner until, at −32°C, and a sucrose concentration of *ca.* 80% w/w, the mixture solidifies into a glass, *i.e.* without crystallising. Compared to the eutectic point, the vitrification point is reached at a supersaturation degree of 43% w/w and a degree of undercooling of 8.4%. The point of maximum freeze-concentration is now commonly referred to as "T_g'". The fact that sucrose does not (cannot) crystallise during freeze-concentration is an indication of its very complex crystal structure, which makes for a low nucleation rate, compared to the imposed rate of cooling. Thus, sucrose will not crystallise in real time.† Instead, it will yield a highly concentrated solid solution in water. This property, which is common to most polyhydroxy compounds (PHCs), is extensively exploited in the formulation of pharmaceutical preparations destined for drying. The important thermophysical characteristics of amorphous PHCs, and their impact on pharmaceutical product quality, will be described in more detail in Chapters 6, 7 and 10.

Knowledge of the kinetics of salt precipitation becomes particularly important for the correct choice of pH buffering mixtures of labile biopharmaceutical preparations during freezing. Returning once again to the crystallisation of phosphate buffer mixtures under equilibrium conditions, it needed to be established whether this behaviour does actually exist in real formulations and under freezing conditions. It was found that the fractions of the two phosphates that do actually precipitate depend on the buffer composition, but not in an additive manner, as well as on the initial buffer concentration and the cooling rate.[56] Where one or the other salt in the buffer mixture precipitates spontaneously during cooling or annealing, the pH of the remaining solution phase will suffer a shift at that point, particularly if the other buffer component is subject to long-term supersaturation. A striking example of this type of behaviour is shown in Figure 5 for a Na_2HPO_4–citric acid buffer (nominally at pH 6.5). The solution was seeded with an ice crystal to initiate freezing at −0.5°C (the ice/sodium phosphate eutectic point). No salt precipitated at

† The expression "real time" is used in this book to denote the observable duration of a physical/chemical process or of an experimental observation, or the shelf life of a dried product. In other words, "real time" will be expressed in a range of 10^1–10^8 s (seconds to years)

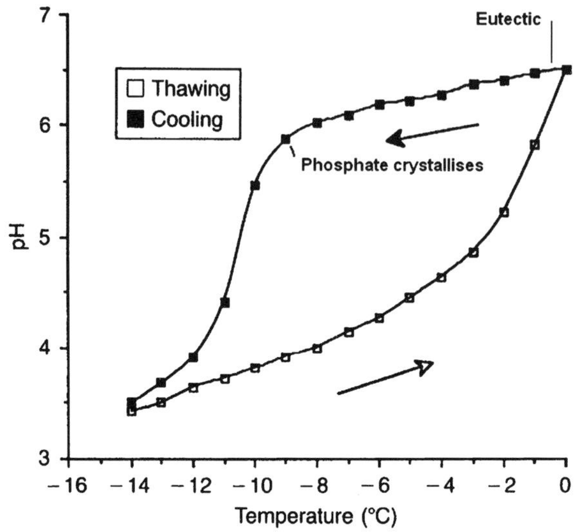

Figure 5 *pH changes during the cooling (and freezing) and rewarming of an ice-seeded citric acid–sodium phosphate buffer mixture, illustrating the effects of delayed crystallisation of sodium phosphate; reproduced from Pikal (unpublished)*

this temperature. During cooling, the solution became increasingly supersaturated until, at *ca.* −10°C, $Na_2HPO_4.12H_2O$ began to precipitate, but slowly. On completion, this produced a major pH shift to 3.5, since only citric acid was left to act as buffer.

5.5 Effects of Freeze-Concentration on Reaction Kinetics

Reaction rates in part-frozen solutions tend to exhibit complex temperature dependencies. According to textbook chemistry, the Arrhenius equation describes the effect of temperature on chemical rate constants. It is therefore sometimes (mistakenly) assumed that reaction rates must decrease during freezing. This slowing down of deleterious reactions is often advanced as one of the benefits of freeze-drying. In fact, the opposite is the case; reaction rates *increase*, sometimes dramatically, during freezing. It is overlooked that the Arrhenius equation applies to homogeneous systems of constant composition, whereas freezing produces a heterogeneous system and continuous changes in concentration, or even composition. Probably because of experimental complexities, few quantitative studies of reaction kinetics in part-frozen solutions are on record, but rate enhancements of several orders of magnitude have been reported, particularly for enzyme-catalysed reactions. One

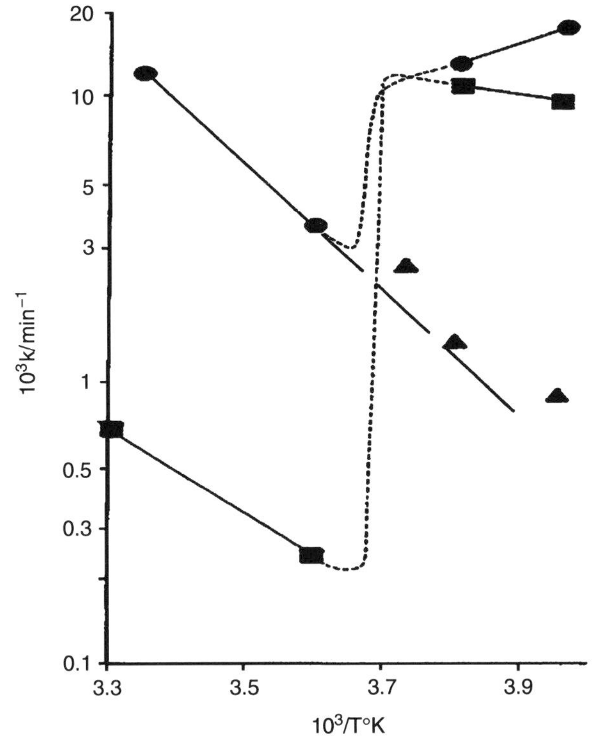

Figure 6 *The oxidation rate of ascorbic acid at low temperatures in undercooled (▲) and frozen solutions: dilute (■) and concentrated (●). Broken lines denote subfreezing temperatures. Redrawn after Hatley et al.*[59]

example, worthy of detailed study, concerns the complex effects of freezing on the acid-catalysed mutarotation kinetics of glucose.[17,58] The observed rate constant initially rises steeply during freeze-concentration; it increases to a maximum at –8°C to decline with further decrease in temperature. The study also shows how any so-called inert additives, *e.g.* salt, significantly affect the mutarotation kinetics during freezing.

For a comparison between the effects of undercooling and freezing on chemical rates, the oxidation kinetics of ascorbic acid by hydrogen peroxide over the temperature range +30 to –20°C, shown in Figure 6, can serve as a good example.[59] It is seen that the effects of cooling, but with the avoidance of freezing, follow the normal Arrhenius type of behaviour. On the other hand, if the solution is allowed to freeze, then the oxidation rate increases during freeze-concentration, probably to assume Arrhenius-type behaviour at some high degree of supersaturation, but probably not with the same activation energy to that observed

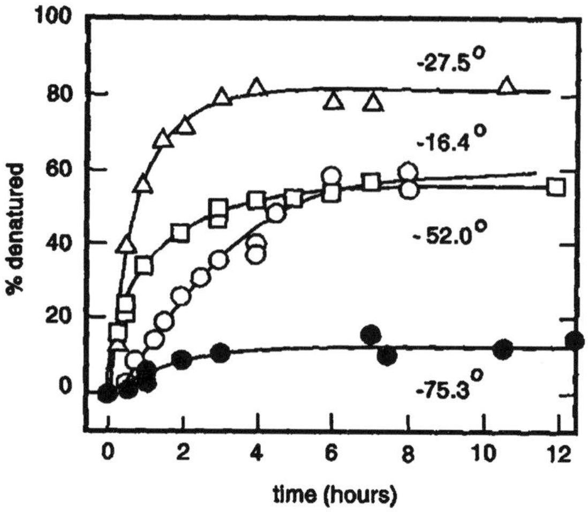

Figure 7 *The inactivation rates of chymotrypsinogen solutions frozen to different temperatures. Modified from Brandts et al.*[61]

for the undercooled solution. Both the dilute and concentrated solutions freeze-concentrate to the same final concentration, but the degree of freeze-concentration produces a much more marked increase in the rate constant of the initially dilute solution. The kinetic rate enhancements that accompany freezing appear to be particularly large in enzyme-catalysed reactions, where rate enhancement in excess of three orders of magnitude has been observed.[60]

The mechanism and kinetics of protein inactivation during freeze-concentration are of particular importance in the context of freeze-drying. Here again, quantitative information is scarce, but the pioneering studies by Brandts, shown in Figure 7 for chymotrypsinogen, illustrate how the rate and degree of inactivation can be affected by subfreezing temperatures.[61] At first sight, the *increase* in the rate, accompanying the *decrease* in temperature, is unexpected. Following the peak at –27.5°C, both rate and extent of inactivation decrease with decreasing temperature. Inactivation is thus affected by temperature in a non-monotonic and non-Arrhenius manner, an observation that is hardly explicable by textbook chemistry. The publication of the data in Figure 7 should have caused a stir and led to the realisation that storage of labile biochemical substances, *e.g.* perishable foods, in the neighbourhood of –20°C is not to be advised. Yet this is still the operating temperature of the typical domestic freezer and also the temperature recommended by most manufacturers for the storage of many of their purified biochemical products! It is indeed striking that the significance of Brandts' findings went completely unnoticed by the scientific community and by the author himself, and especially by those who were at that time considered to be experts in the discipline of cryobiology. It was not until the importance of amorphous states, particularly glasses, in the life sciences came to be realised, that Brandts' cryoinactivation results began to make sense.[17]

5.6 Complex Ternary and Multicomponent Phase Behaviour

Although the majority of freeze-dried pharmaceutical preparations consist of complex mixtures of several often chemically quite different components, it is surprising that, until fairly recently, their phase properties were not taken into account during either product or process development. The following discussion focuses mainly on the phase behaviour of particular model mixtures in pharmaceutical and biopharmaceutical freeze-drying practice. We feel this generalisation from the particular model to complex formulations to be permissible, because in many practical situations the preparation to be dried will consist mainly of water and excipient(s). In many cases, the actual concentration of the bioactive drug component is then such that it can be neglected, as far as the freeze-drying process is concerned.

A distinction is now introduced between primary and secondary physical processes. Thus, *primary* crystallisation refers to the process that takes place during cooling; it may or may not lead to a complete crystallisation. *Secondary* crystallisation refers to the process that takes place during annealing, thermal cycling or reheating. It is frequently the completion of a phase separation that should, under equilibrium conditions, have taken place completely during cooling.

Consider a ternary aqueous mixture: as cooling proceeds, three general types of phase behaviour can be distinguished, according to the properties of the residual solution after initial (primary) ice crystallisation:

(i) The freeze-concentrate may form a kinetically stable (but thermodynamically unstable) amorphous phase, capable of eventual vitrification.
(ii) The freeze-concentrated solution may form a "doubly unstable" glass[62] in which a partial ice/solute crystallisation may occur during annealing, thermal cycling and/or drying.
(iii) A binary ice/solute (eutectic) crystallisation may occur after primary ice crystallisation, with the residual solution forming a kinetically stable glass.

A complete crystallisation of ice and all the solute(s) in real time is rare. Nevertheless, the physical and chemical properties of the freeze-dried product depend critically on the equilibrium phase behaviour, as defined by the phase rule, and also on any transient non-equilibrium excursions, induced by the particular processing conditions employed during freezing and drying. The system water–sucrose–NaCl serves as a good illustration of these three possibilities.

Multicomponent systems can usually be considered as either pseudo-binary, when only water crystallises, or pseudo-ternary, when water and a solute crystallise. When ice is the only crystallising species, the process can be adequately described with the help of a simple two-dimensional temperature–composition phase diagram, as shown in Figure 1a in Chapter 4. To describe in full the crystallisation and phase relationships in the ternary system, a three-dimensional phase diagram is required, but as a simplification, a triangular projection on the composition base can usually be employed. The following is a summary outline of the construction and interpretation for a model system of pharmaceutical significance, water–NaCl–sucrose, where the data were obtained from an in-depth study by DSC methods.[172]

An analysis of the experimental data was performed by consideration of the three binary systems: water–NaCl, water–sucrose and NaCl–sucrose. Of these, the first system has been thoroughly studied, and the transition temperatures and compositions are included in Table 2. The situation is more complicated for water–sucrose mixtures, mainly because any crystallisation processes of sucrose or any of its hydrates from aqueous solutions are likely to be very slow, perhaps impossible to determine by DSC methods in real time. Some hydrates have, however, been identified by X-ray and chemical analytical methods; they are also included in Table 2 and the (probable) equilibrium phase diagram is shown in Figure 8. As regards the anhydrous system NaCl–sucrose, no quantitative information exists. The crystallisation of NaCl from its solid solution in amorphous

Table 2 *Invariant points in the binary systems water–sucrose and water–NaCl*

Equilibrium	*Composition (%*	*T (K)*	*System*
Eutectic L/ice + I	44.1	263.3	W–S
Peritectic L + II/I	36.3	274.0	
Peritectic L + S/II	35.3	275.8	
Eutectic L/ice + III	76.9	252.15	W–NaCl
Peritectic L + NaCl/II	73.7		273.28

L, liquid; S, sucrose and W, water.
Roman numerals represent crystalline hydrate phases: I = S.3.5W; II = S.2.5W; III = NaCl.2W.Adapted from Shalaev and Franks[172]; see also Figure 8.

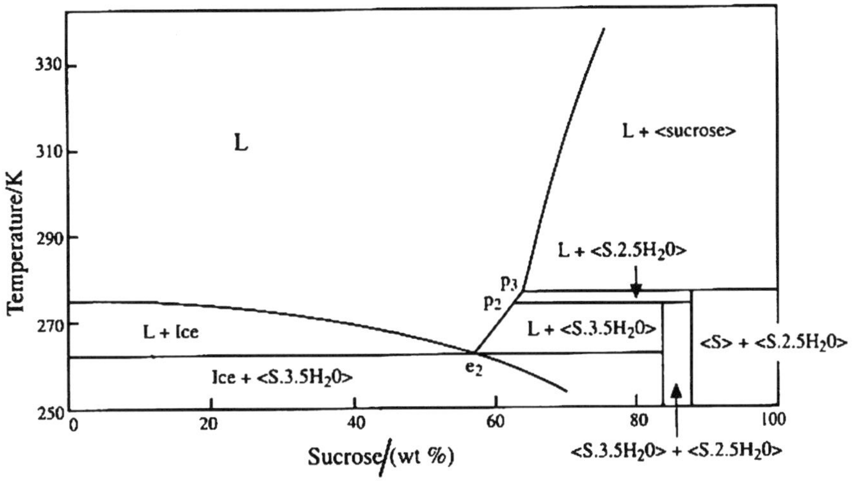

Figure 8 *The sucrose–water solid/liquid phase diagram. S, sucrose; L, liquid solution; e, eutectic; p, peritectic; and< >, crystalline phases. Reproduced from*[172]

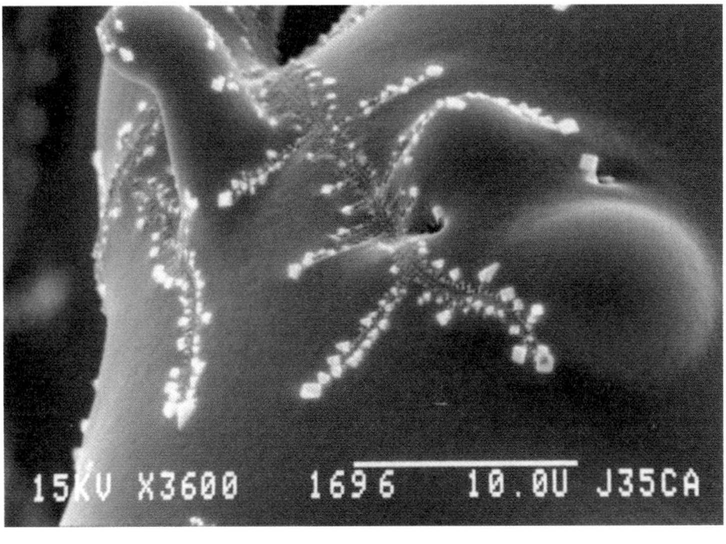

Figure 9 *Scanning electron micrograph of a freeze-dried solution, containing sucrose and NaCl at a mass ratio of 6:1, after warming for a short period to a temperature slightly above T_g, resulting in the crystallisation of NaCl from the glassy sucrose solvent matrix. Preferred crystallisation sites are seen to exist at the matrix cracks. Bar line corresponds to 10 μm. Franks & van den Berg*[63]

sucrose has, however, been observed by electron microscopy, and a representative picture is shown in Figure 9, where the growth of well-defined cubic NaCl crystals of up to *ca.*100 nm dimensions from its solid solution in amorphous sucrose solvent is clearly illustrated.[63,64]

A direct study of phase relationships in ternary systems of these components presents severe challenges, mainly because of the very slow crystallisation of sucrose. Reliable experimental studies are thus limited to the water-rich domain of the phase diagram. A combination of old solubility, refractive index and X-ray data, with the recent DSC results has provided semi-quantitative estimates of several equilibrium phase transition points (eutectic and peritectic), indicating also coexistence curves between pairs of the following six components: water, NaCl, $NaCl.2H_2O$, S, $S.2.5H_2O$ and $S.3.5H_2O$, where S stands for sucrose. All this information is important in the context of freeze-drying because the optimum freeze-drying cycle and the processing details, and also the structure and stability of the dried product depend on such transition points. Although the actual numbers (temperatures, compositions, phases) are seldom known, the product and process developers should be aware of complex phase relationships that are the determinants of the product behaviour during and after freeze-drying.

Another ternary model system, water–NaCl–mannitol, is of equal importance for freeze-drying operations, mainly because D-mannitol, an acyclic alditol, $CH_2OH(CHOH)_4CH_2OH$, enjoys great popularity as an excipient among formulation developers. Its phase complexities are of a different kind from those encountered with sucrose, because of all the PHCs employed in pharmaceutical technology, mannitol is uniquely able to crystallise from freezing solutions in real time. In addition, the alditol can also exist in several polymorphic forms (α, β and δ), and under certain processing conditions, it can crystallise as a monohydrate, already referred to earlier. The β-form is the commercially available polymorph, while it is claimed that the δ-polymorph is preferable as a tabletting excipient because of its superior response to compaction. The complex crystallisation behaviour of mannitol has long been known,[65] but even now, and despite many attempts, the conditions that favour the precipitation of a given polymorph are not yet fully understood. It appears also that mixtures of polymorphs can co-crystallise, a phenomenon that has led to spurious claims about numerous "new" polymorphs, leading to the opinion that "Different names of the crystal forms by several authors have left behind a chaotic picture in the literature about which physicochemical properties belong to which crystal form".[66] Nevertheless, the physical properties of the three "clean" polymorphs are now (probably) adequately well established, and a selection is shown in Table 3. Turning now to the other binary system, mannitol–NaCl, an unexpectedly high melt miscibility has been found (up to 7.5% by weight of NaCl) at the eutectic temperatures, 150°C for the α- and β-polymorphs, and 139°C for the δ-polymorph.[67]

Table 3 *Polymorphism and selected physical properties of mannitol*

Modification	*I*	*II*	*III*
Crystal habit	Prismatic rods	Prismatic rods	Needles
Melting point (°C)	166.5	166	155 (incongruent)
Enthalpy of fusion ($kJ\ mol^{-1}$)	53.5	52.1	53.7
Density (gcm^{-3})	1.490	1.468	1.499 ± 0.004
Specific heat at 25°C ($Jg^{-1}\ k^{-1}$)	1.383 ± 0.009	1.273 ± 0.008	1.263 ± 0.002

Source: Data from Burger *et al.*[66]

Table 4 *Crystallisation of glycine from frozen solutions*

Substance	*Additives*	*pH*	*Crystallisation*	*Eutectic (°C)*	*Comments*
Glycine	None	7	Rapid	–3.4	B-polymorph
Glycine-HCl		7	Very slow	–28	
Glycine-Na	NaCl		Slow	–17.8	$T_g \approx -80$
Glycine		3	Slow, γ-polymorph	–3.5	$T_g \approx -70$

Source: Data from Akers *et al.*[68]

Similar results were obtained by substituting KCl for NaCl, but LiCl produced a lower eutectic point and also inhibited the crystallisation of mannitol. The earlier paragraphs may suggest to the product developer that, even if only for regulatory reasons, the use of mannitol in pharmaceutical preparations should be avoided, a valid conclusion in the author's own experience.

Amino acid crystallisation from a freezing aqueous solution provides yet another interesting facet of the complexities that can be encountered during freeze-drying operations, and that therefore deserves mention. Here, chemically binary systems, mixtures of water + amino acid display complexities, arising from the zwitterionic nature of amino acids (and peptides). Taking glycine as an example, its crystallisation behaviour is strongly influenced by pH and salts;[68] examples are shown in Table 4.

5.7 Supersaturation and Vitrification: The Importance of the State Diagram

Having already established that the phenomenon of supersaturation (undercooling) of a solution during the removal of water may be transient, despite being a common event, an important question then arises: what might be the lifetimes of supersaturated solutions under freeze-drying conditions, before solute crystallisation (precipitation) occurs spontaneously and goes to completion? For most PHCs, these lifetimes are likely to be extremely long, so that few experimental data for PHC crystallisation from part-frozen solutions exist. Notable exceptions are mannitol and lactose. The former can crystallise quite rapidly, even at subfreezing temperatures, while the latter is known to crystallise, often inadvertently, as in ice cream, even when stored at –20°C.

The previous section dealt with *equilibrium* phase relationships during freezing and drying, and it was shown that even they can be complex. We now extend the discussions to *non-equilibrium* situations, such as supersaturation, a common event, which has already been touched upon. Where lifetimes of supersaturated solutions extend into years or centuries, it becomes permissible to treat such mixtures as pseudo-equilibrium solutions. Consider the case of a sucrose solution that is being cooled at a constant rate. The phase rule predicts that at some temperature, the eutectic point, sucrose should crystallise. However, nucleation probabilities of PHCs, especially from aqueous solutions, are generally very low. This arises from their complex crystal structures. As in ice, PHC molecules in the crystals are linked by weak hydrogen bonds between their many –OH groups to form infinite three-dimensional chains and networks.[69,70] Furthermore, interactions between sugar –OH groups and those between sugar –OH groups and water molecules are closely similar in energy and configuration. For these reasons, most PHCs are reluctant to crystallise, so that on cooling, even at quite low rates, ice will continue to grow beyond the (notional) eutectic point and the solution becomes increasingly supersaturated and viscous. The increasing viscosity leads to a slowing down of ice crystallisation, until at some characteristic temperature no further freezing can be detected in real time. The solution is then said to have reached its glass transition temperature, usually denoted by T_g', the temperature of maximum freeze-concentration. This temperature is of the utmost importance for the development of rational and effective freeze-drying cycles. Its dependence on the solution composition can be incorporated into the solid–liquid phase diagram, but with the proviso that the glass transition profile does *not* represent a solid–liquid phase coexistence curve and is purely an isoviscosity profile; *i.e.* it connects the T_g values of mixtures that have the same viscosity, often taken as $\approx 10^{14}$ Pa s. This representation of equilibrium phase data with kinetic (vitrification) data was in the past described as a "supplemented phase diagram",[42] but it is nowadays more commonly referred to as a "state diagram".[71] Such a diagram is shown in Figure 10 for the system water–sucrose.

For many aqueous systems, especially PHC solutions, the actual positions of their true eutectic points are quite uncertain, giving rise to significant discrepancies in reported solubility data by different workers.

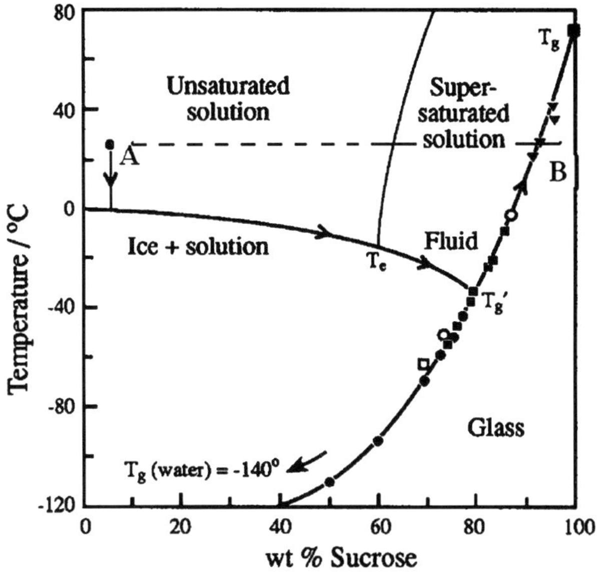

Figure 10 *Solid/liquid state diagram of the sucrose–water system showing also the glass transition profile. Symbols correspond to data taken from diverse sources*

A particularly striking example is provided by Miller *et al.*[72] for α,α-trehalose. Their long-term equilibration experiments led to a eutectic point of –2.5°C (29.8% w/w), in reasonable agreement with a value of –4.4°C,[73] but differing substantially from –18.8°C, earlier reported.[74] Such uncertainties in the correct equilibrium properties inevitably produce even larger uncertainties in the correct value of T_g' and C_g', the unfrozen water content at maximal freeze-concentration. Miller *et al.* cite recent literature values of T_g', ranging from –22 to –35°C, to demonstrate the large uncertainties in this number, which is of unique importance in the calculation of primary drying times (ice sublimation); see Chapter 8. Of almost equal importance is C_g', the independent determination of which is also beset by considerable experimental problems (see Chapter 9).

From the earlier definition of the glass transition, it might be expected that below T_g', crystallisation is inhibited, and this is generally the case, at least for practical time scales. There are, however, reports describing crystallisation of polymers in real time at temperatures well below the glass transition.[75] It must of course be remembered that, on a molecular basis, glasses are undercooled *liquids*, and they differ from ordinary liquids mainly in the rates of their transport properties. These differences can be usefully exploited in the formulation of pharmaceutical preparations, as will presently be discussed.

5.8 Aqueous Glasses as Solid Solutions

It is now clear that, on a molecular scale, glasses must be treated as liquids, because they do not possess long-range structural order. Their observed quasi-solid properties originate from their extremely high viscosities, estimated to lie in the range of 10^{12}–10^{14} Pa s at the glass transition temperature T_g. The magnitude of such viscosities can be better appreciated by considering the inverse property, the rate of viscous flow. For glasses at T_g, linear flow rates are then measured in μm $year^{-1}$ or mm $century^{-1}$. Though the study of inorganic glasses has a long history,[76] the exact nature of the glass "transition", especially that of water, has long been, and still is, something of a mystery.[77] However, for practical purposes, pharmaceutical glasses can be regarded as solids in pseudo-equilibrium and treated as equilibrium "phases". This approach greatly facilitates their study by thermodynamic and structural techniques. The complexities arise mainly from their kinetic properties, and they will be discussed in the following chapter.

Reference was made in the preceding chapter to the transient nature of supersaturated (undercooled) states. From a freeze-drying perspective, a glassy excipient must possess a lifetime of at least the duration of the drying process. It must also have T_e and T_g' values that can be accommodated within the proposed drying cycle. Eutectic points vary over a large range of temperatures and compositions. For instance, the T_e of Na_2HPO_4 is –1°C, whereas that of $CaCl_2$ is –52°C. Where a solution is formulated with the aim of producing a crystalline freeze-dried product, only those excipients with a high eutectic temperature, *i.e.* $T_e > -30$°C, and an acceptably high rate of nucleation would find practical application. Similarly, if the aim is to produce an amorphous product, excipients that do not readily crystallise, but give rise to kinetically stable supersaturated solutions would be preferred, especially those with high T_g' values, say > -30°C. This limitation makes for the exclusion of most mineral salts. Table 5 presents a selection of eutectic and glass transition data to illustrate that point. Given that eutectic points of water-soluble substances cover the range of –1 (maleic acid) to –88°C ($ZnCl_2$), and T_g' values the range –33 (glutamic acid) to –111°C ($MgCl_2$), it remains an open question whether there is a causal relationship between a crystal structure, the solubility and a corresponding T_g' value.

Thermodynamically, a multicomponent glass can usually be treated as a homogeneous solid solution. It might then be expected that for chemically similar molecules the conventional mixing rules might apply, perhaps even be reflected in the T_g–composition relationship. This has indeed been found to be the case for binary PHC blends, where neither

Table 5 *A selection of salts that form stable eutectics (E), unstable glasses (UG) or that remain amorphous (SG) during freezing*

Salt	T_e *(°C)*	T_g' *(°C)*	*Class*
$Na_2HPO_4.12H_2O$	−1		E
Na_2SO_4	−2		E
Na_2CO_3	−3		E
KCl	−11		E
Urea	−13		E
$(NH_4)_2SO_4$	−19		E
Glutamic acid	−0.1	−33	UG
NaCl	−23	??	UG
$K_2HPO_4.6H_2O$	−14	−81	UG
Tris–HCl	−13	−86	UG
Guanidine–HCl	−24	−103	UG
Na citrate		−42	SG
NaH_2PO_4		−65	SG
K citrate		−100	SG

component is prone to crystallisation from the anhydrous mixture in real time.[78] It is interesting to note that the glass transition/composition profile of sucrose does not resemble that displayed by an equimolar mixture of glucose and fructose. For chemically dissimilar components, and temperatures above T_g, the molecular mobilities of the various components in real time increase rapidly with a rise in temperature. Phase separation then becomes a distinct possibility. Figure 9 shows such behaviour for a freeze-dried solution, originally containing water, sucrose and NaCl. After the almost complete removal of water, the system became a solid solution of NaCl in sucrose solvent and had a very low residual water content. The figure shows the state of the solution after it had been heated to slightly above its T_g and kept at that temperature for several minutes, after which it was quenched back to its original low temperature.[63] Phase separation, in the form of the growth of NaCl (or $NaCl.2H_2O$) as well-formed crystals of submicrometre dimensions, embedded in the glassy solvent (sucrose), is clearly indicated. Depending on the temperature, the rate of crystal growth can be closely controlled.

As regards structural properties of PHC glasses that find applications in the formulation of drugs destined for freeze-drying, hard information is limited. By "hard information" we understand diffraction data from which atomic distances and molecular orientations in space can be calculated. While there is an abundance of X-ray diffraction data for crystalline PHCs, such techniques do not "see" hydrogen atom positions. No direct information about hydrogen bonds can therefore be obtained, except by speculation, based on observed O–O distances.

Since the crystal structures of water and PHCs are characterised by chains, rings and networks of molecules held together (or apart) by more or less tetrahedrally arranged and more or less linear O–H–O hydrogen bonds (sp^3 hybridisation of orbitals), X-rays are of limited value. Added to this limitation is the requirement for a long-range order, necessitating a certain minimum number of unit cells. Amorphous phases do not possess such long-range structures, making data processing of diffraction data less informative, because the substance under test must be treated as a liquid. Thus, whereas single crystal diffraction data provide spatial coordinates (of carbon and oxygen atoms), measurements on liquids only provide statistical probabilities of such coordinates. No information is provided from which angular coordinates (configuration of molecules) can be calculated with any degree of confidence.

The situation has been vastly improved since the advent of neutron scattering because neutrons can "see" or be made to "see" hydrogen atoms. It has thus become possible to "observe" hydrogen bonds and the configurational arrangement of O–H groups in space. The method has also been applied to aqueous solutions, and much valuable information about the nature of the so-called hydration shells has become available.[33] Efforts to gain a better understanding of differences between PHC crystals and glasses are still continuing, but a comparative neutron scattering study of glucose in the crystalline, the liquid and the glassy state has revealed a marked similarity between the intermolecular distances and configurations of bonds in the crystal and the amorphous states.[79] Thus, the glass can be regarded, at least structurally, as a distorted crystal, with distributions of O–O distances and O–H–O angles. An important difference is that the density and detailed molecular configuration of the glassy state depend on the method of preparation. Other, but important, marked differences between the two states show up in their kinetic and transport properties.

At a very fundamental level, PHC–PHC interactions and PHC hydration can be studied by Monte Carlo (MC) and molecular dynamics (MD) computer simulations. Because of enantiomeric and configurational complexities of molecules with the common formula $C(H_2O)_n$, where commonly $5 < n < 12$, such studies require some drastic approximations. Nevertheless, interesting results have been reported, for instance on the respective crystal, fused and glass structures of glucose[80] and on trehalose hydration.[81] At an even more basic level, some marked differences in PHC –OH group interactions with water have been reported for PHCs that are isomeric (*e.g.* glucose, mannose and fructose) and for the individual –OH groups within the same molecule with water.[82] At present, non-agreement between results reported by different workers appears to be due mainly to the choice of water potential functions and approximations in the permitted motional (rotation, libration) degrees of freedom within the PHC molecules. It must also be borne in mind that any agreement between computed and known experimental properties is in itself no evidence for the correctness of the simulation.

A characteristic property of ternary and multicomponent aqueous solutions, especially where at least one of the components is a polymer, is the occurrence of demixing, resulting in the appearance of two aqueous liquid phases in equilibrium. This phenomenon finds practical application as liquid/liquid extraction in the downstream processing of proteins. Quite remarkable, however, is the coalescence of such coexisting phases into a homogeneous solution upon *cooling*. This phenomenon, known as "lower critical demixing", is related to that of cold inactivation of proteins, already discussed earlier, see also Figure 1,[49] but similar behaviour is observed for mixtures of some small molecules in aqueous solutions. By extrapolating from the mobile liquid to the immobile liquid, the glass, it is not inconceivable that a similar process, the separation of a homogeneous solution into two coexisting phases, might occur during freezing, so that the separated phases would then persist in the vitrified system. Indeed, the existence of two amorphous phases in frozen and freeze-dried solutions has been reported;[47] the phenomenon has been given the name of *polyamorphism*, but it is still much debated in the literature[83]. Its possible impact on freeze-drying operations is clear, especially because vial-to-vial differences in texture and reconstitution behaviour cannot be ruled out where the possibility of random generation of polyamorphism exists.

These examples of physical and chemical changes that may accompany the removal of water by freezing will serve to illustrate that the freezing of multicomponent solutions, such as are common in pharmaceutical formulations, is by no means a simple affair. The removal of water should be thoroughly investigated and understood in terms of phase changes, mutual miscibilities, non-equilibrium states and slow kinetics that might help efforts to characterise the freeze-concentrate. Only by such a better understanding can the product and process development stages be refined to yield high-quality products by economic processing.

CHAPTER 6

Physical Properties of Crystalline and Amorphous Solids

6.1 Crystalline and Amorphous Solids: A Comparison

The word "solid" is in everyday parlance applied to those forms of matter that, on a macroscopic scale, tend to retain their shape and volume, and possess some degree of rigidity. On a molecular scale, however, a distinction must also be drawn between two very different types of solids. The crystalline state, in which the atoms or molecules are arranged in a regular and precise spatial array, is the familiar solid form, characterised by its long-range order and thermodynamic stability. For completeness, it must be emphasised that certain crystalline types, obtained by cooling from the fused state, may not necessarily result in the most stable form, *i.e.* the form of lowest energy. Some elements (*e.g.* phosphorus and sulfur) and many chemical compounds can exist in a variety of crystalline states (polymorphs), each of which is stable only within a well-defined domain in phase space. Outside that domain, they are metastable and liable to spontaneous solid/solid transitions, resulting in a more stable form. Ice is a well-known example of complex polymorphic behaviour. Up to the present, 12 distinct polymorphs, some of them metastable, have been prepared and characterised by X-ray and/or neutron diffraction;[84] some of them are included in Figure 9 in Chapter 4. There is no reason why additional polymorphs may yet be discovered, because even "ordinary" ice (hexagonal ice) is not the form of lowest energy; the statistical distribution of protons gives rise to the well-known residual entropy, first described by Linus Pauling.[85]

Another type of crystal diversity, somewhat related to polymorphism, is brought about by the effects of temperature and/or pressure on chemical composition. The phenomenon is common with solvates, and particularly hydrates, that are formed by many ionic and molecular compounds, when they are crystallised from aqueous solution. Common examples include Na_2SO_4, Na_2CO_3 and alkali metal phosphates, most of which can exist in a variety of hydrated crystalline forms. Organic hydrate formers, of high relevance to freezing include glucose.H_2O, maltose.H_2O, lactose.H_2O, α,α-trehalose.$2H_2O$, β,β-trehalose.$4H_2O$ and raffinose.$5H_2O$. It should be clearly understood that any statement in the literature that a particular carbohydrate does not form hydrates must be viewed with scepticism. It can only mean that no such hydrate has yet been detected. It is likely that many "hidden" sugar hydrates could exist under appropriate, perhaps extreme, conditions. The recent discovery of a previously unknown hydrate of mannitol, a favourite pharmaceutical excipient, gave rise to freeze-drying problems, until the nature of the problem was recognised: water was released into the freeze-dried product during the decomposition of the mannitol hydrate into one of its well-known anhydrous polymorphs.[57]

Other materials that also appear to be solid may, however, be devoid of such long-range atomic/molecular order. Although some short-range order is likely to exist, such materials are classified as "amorphous". There is no general agreement about the dividing line between crystallinity and amorphism, *i.e.* between long- and short-range order, or how many unit cells are required for a material to display the physical characteristics of crystallinity.[173] In principle, most materials can exist in both forms, depending on the method of preparation, processing and storage conditions. Sucrose, for example, is commercially supplied as an anhydrous crystalline solid at ambient temperature. If, however, crystalline sucrose is melted and then rapidly cooled, crystal nucleation does not occur spontaneously at the equilibrium melting point. Instead, the undercooled liquid sucrose becomes increasingly viscous until, over a particular range of temperature, the viscosity suddenly increases by several orders of magnitude and the sucrose solidifies into a *glass* (*i.e.* an amorphous solid).

An amorphous "solid" is therefore characterised by its lack of a long-range ionic, atomic or molecular order. It is thus an undercooled liquid that is thermodynamically unstable with respect to the crystal form, but the energy barrier to viscous flow is high enough to prevent its reversion to the stable, crystalline state, at least *within the normal time scale of observation.* In a similar manner, other kinetic rate processes, *e.g.* chemical reactions, are severely inhibited in the glassy state. This makes the glass a valuable stabilising matrix for chemically labile materials.

The multiple possible energy states can be illustrated by means of a so-called landscape, as shown in Figure 1, and the reader is referred to a recent excellent summary primer of undercooled aqueous liquids and vitreous aqueous states.[86]

The existence, and especially the coexistence of dual solid states, crystalline and amorphous, gives rise to certain conceptual problems, when viewed in terms of "textbook chemistry". Thus, students are taught that thermodynamics describes energy relationships between different states of matter but provides no direct information about the kinetics or mechanisms of transformations. On the other hand, materials science and technology teach that the (thermodynamically) unstable states of materials are generally much more "useful" than their stable counterparts. A few examples will drive this point home: the stable equilibrium state of a mixture of oil and water consists of two almost immiscible liquid layers. Mayonnaise, on the other hand, is an unstable mixture of oil and water in the form of an emulsion. It obtains its apparent stability from clever formulating and processing. "Solder" is a generic term for metallic mixtures, termed "alloys", *e.g.* zinc and copper, or tin and lead. All alloys are thermodynamically unstable because the two metals do not form single crystalline phases, but they can be made to mix in the fused state, from which the solder is obtained by quenching, *i.e.* as an undercooled, viscous fluid with some of the thermomechanical properties of a plastic solid. An equivalent mixture of the two crystalline metals would be technically useless for applications in which the low softening point is

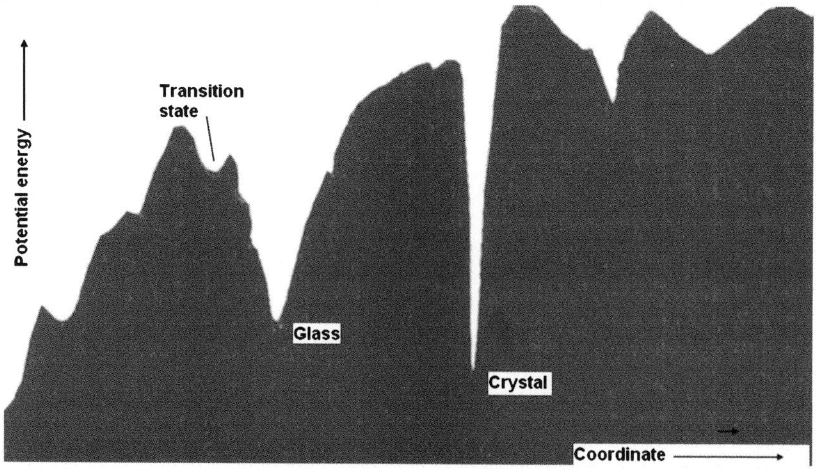

Figure 1 *A configurational "landscape" for an amorphous substance. Once caught in a deep potential well, the system requires a high degree of energy of activation to escape and undergo the necessary changes for crystallisation*

the desired property; it can only be achieved by processing. In other words, compounding and processing of inherently immiscible materials produce an apparent miscibility that provides technologically desirable properties. This does not in any way offend against the laws of thermodynamics, which state that *under equilibrium conditions* the two materials are immiscible. What the technologist has learnt is how to delay the reversion to equilibrium (in these cases, a demixing) by the introduction of kinetic energy barriers. "Shelf lives" of such thermodynamically unstable states can and do range over many orders of magnitude, and "stability" is only apparent and depends on the period of observation.

A graphical illustration of the crystal/glass dichotomy is shown by the photomicrographs of raffinose in Figure 2. The left picture was obtained from a crystal of raffinose.5H_2O, the stable form; it shows a high degree of birefringence, characteristic of crystals. The picture on the right displays the same shape and dimensions of the original crystal, but it has been obtained after the complete vacuum dehydration of the crystal, and it retains no vestige of the original birefringence. The material is thus amorphous (glassy) on the molecular scale, *i.e.* the long-range order of the hydrogen-bonded network has been destroyed. The apparent retention of the original crystal shape results from the high viscosity of the glassy form. A superficial microscopic examination might well have led to the erroneous conclusion that the anhydrous material was crystalline.

Figure 2 *Micrographs of a raffinose crystal (left), showing pronounced birefringence, and the same crystal (right) after dehydration by vacuum treatment at room temperature. The ultrastructure resembles that of the crystal, but the absence of birefringence is evidence of the amorphous state; from Kajiwara and Echlin, Pafra Biopreservation, unpublished*

The bottom line of this discussion is therefore that a study of the science and technology of amorphous solids requires the introduction of a kinetic element into classical thermodynamics.[87] Also, in practice, the importance of kinetics far outweighs the predictions of the equilibrium phase rule, because equilibrium might be hours, years, centuries or millennia away.

6.2 Non-equilibrium Processes in Amorphous Solids

For kinetic processes that are slow, relative to the period of observation, we need to be aware of the time evolution of equilibrium properties (energy, volume, structure, solubility, *etc.*). Consider the reaction between oxygen and hydrogen gases at 25°C. According to textbook thermodynamics, the gases should react to form liquid water because at that temperature water has a lower free energy than the mixture of the two gases. However, experience teaches that the reaction does not take place, at least not within a measurable period of observation, because the kinetic energy barrier is very high. It requires the addition of a catalyst, *e.g.* MnO_2, to cause the reaction to take place spontaneously and explosively!

The *average* structure in an amorphous (disordered) system is constant with time but subject to spatial fluctuations (dynamic equilibrium). Where such fluctuations involve the breaking and remaking of bonds or spatial reorganisations of molecules or groups of molecules, relaxation rates towards equilibrium may become very low (relative to the period of observation). Such a system can then be considered as petrified on the experimental time scale and take on some of the thermal and mechanical properties of a crystalline solid. All thermodynamic properties will, however, remain partially time dependent. If such a system is subjected to a sudden perturbation, *e.g.* a pressure or temperature jump, one observes an immediate response, followed by a slow approach (relaxation) to quasi-equilibrium.[88] The process of structural relaxation is shown schematically in Figure 3 for a fused system that has been subjected to a temperature jump (quenching). The initial, rapid response is due to an elastic, crystal-like increase in the density, but without concomitant changes in the bonding pattern. The following, slow relaxation process involves a rearrangement of bonds between atoms or molecules, to yield a structure of lower energy that may more closely resemble the equilibrium structure. In the case of amorphous polyhydroxy compounds (PHCs), such relaxations involve rearrangements of intermolecular hydrogen bonding patterns, until a new structure with a lower energy has been achieved.

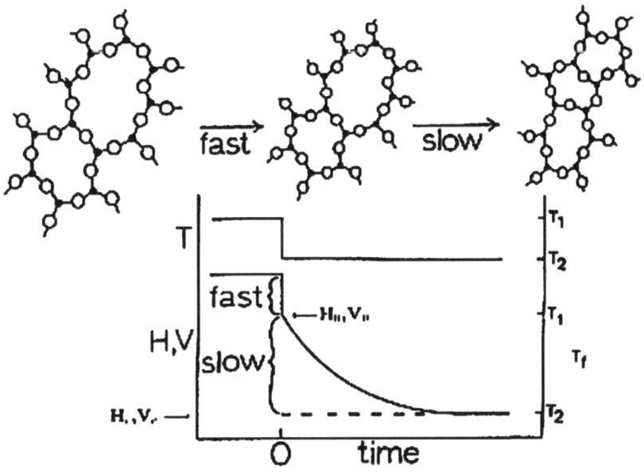

Figure 3 *Changes in enthalpy or volume with time during an isothermal structural relaxation following a temperature jump. Reproduced with permission from Moynihan*[88]

For a temperature or pressure jump, the rate of the structural relaxation is expressed in terms of a characteristic *relaxation time* τ. As a working approximation, the rate of approach of the density and configuration to their equilibrium values can be treated as first-order kinetic processes. Thus, the change in enthalpy accompanying the relaxation is expressed by

$$d(H_t - H_e)/dt = -k(H_t - H_e) \quad (1)$$

where H_t is the enthalpy after time t, H_e is the equilibrium enthalpy and k is the first-order kinetic rate constant.

On integration of Equation (1), a relaxation function is obtained as

$$\phi(t) = (H_t - H_e)/(H_o - H_e) = \exp(-kt) = \exp(-t/\tau) \quad (2)$$

where $\phi(t)$ changes from unity (at $t = 0$) to zero (at equilibrium), H_o being the initial enthalpy, at $t = 0$.

A more refined and realistic treatment takes account of multiple, partly superimposed relaxation processes (stretched relaxation) by including a "distribution parameter" β in the relaxation function, where $0 < \beta < 1$:

$$\phi(t) = \exp(-t/\tau)^{\beta} \quad (3)$$

A second approximation stipulates that the temperature dependence of τ follows Arrhenius kinetics, so that

$$\tau = \tau_o \exp(\Delta H^*/RT) \quad (4)$$

where ΔH^* is the activation energy of the relaxation process. It has been found that in practice ΔH^* is equal to the energy of activation of viscous flow.

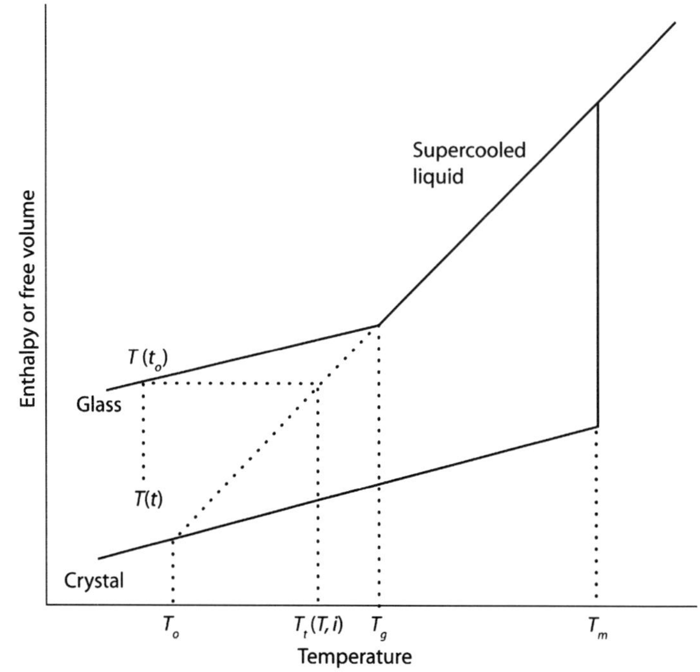

Figure 4 *Relationships between liquids, crystals and glasses. T_m: equilibrium melting point; T_g: glass transition at a given cooling rate; $T(t_o)$: glass transition at a given time from which annealing (ageing) is to be measured; $T(t)$: glass transition after time t; $T_f(T,t)$: "fictive" temperature and T_o: Kauzmann temperature or isentropic temperature. For details, see text and Chapter 11*

The time dependence of some thermodynamic properties of amorphous materials *below* the glass transition is shown in Figure 4. The actual position of T_g depends on the cooling rate. For many practical purposes, energies and densities, along with most other properties of vitrified materials, are expressed in terms of their relaxation times. A fictive temperature $T_f(T,t)$, is introduced, which represents the temperature at which the quasi-equilibrium enthalpy of the undercooled liquid equals that of the non-equilibrium glass; $T_f(T,t)$ then accounts for the temperature and time dependence of the non-equilibrium state. The significance and usefulness of the fictive temperature is discussed in Chapter 11.

6.3 Slow Relaxation during Cooling and Heating

Figure 5a and b represent the thermal or mechanical responses of a fluid that is being subjected to successive temperature jumps ΔT and allowed to recover after each step for fixed time periods of Δt. It is seen that after the first step, the resting time is sufficiently long to enable the material to recover to its previous state. After the second step, a lag is already apparent that increases at the third step. At the fourth step, the material begins to behave as a solid, because no recovery is observed during the resting interval.

Considering now the responses when the material is reheated (marked by crosses in Figure 5b) under the same conditions, the apparent solid state persists until the second temperature jump, when the material first begins to relax, but from its previous vitrified state, which had been reached after the third cooling step. Thereafter relaxation is as expected, until the resting time Δt is once again sufficient for equilibrium to be attained. This description and an inspection of Figures 4 and 5 demonstrate that the responses of the material during cooling and heating are *not* identical. The intermediate temperature region in which the system passes from the fluid to a glassy response is identified with the *glass transition region* (where $\Delta t \approx \tau$). In this region, the $H(T)$ and $V(T)$ curves for cooling differ from the corresponding heating curves. The glass transition region is thus seen to be cooling/heating rate dependent, *i.e.* it depends on $(\mathrm{d}T/\mathrm{d}t)$. Whereas enthalpy and volume change monotonically with temperature, their temperature derivatives (specific heat and coefficient of expansion) display a sigmoid temperature dependence. This point is important for the experimental determination of glass temperatures and relaxation rates.

6.4 Glass Transition: A Summary

In light of this discussion, the glass transition can now be described as a narrow temperature range over which the specific heat and thermal expansion coefficient of amorphous materials undergo more or less sudden changes due to relaxation effects. In laboratory practice, recovery times might lie in the range of 10 min to 1 h. It is thus important to realise that the glass transition is *not* a first-order phase transition in the sense of crystallisation, melting or polymorphism, but a kinetic phenomenon that extends over a narrow temperature range and is dependent on heating/cooling rates and the thermal history of a material. A corollary is that the density of a glass depends on the temperature at which it was formed, and also on the time for which it was stored (annealed). Some practical aspects of relaxation phenomena as they relate to the behaviour of freeze-dried pharmaceutical products will be further explored in Chapter 11.

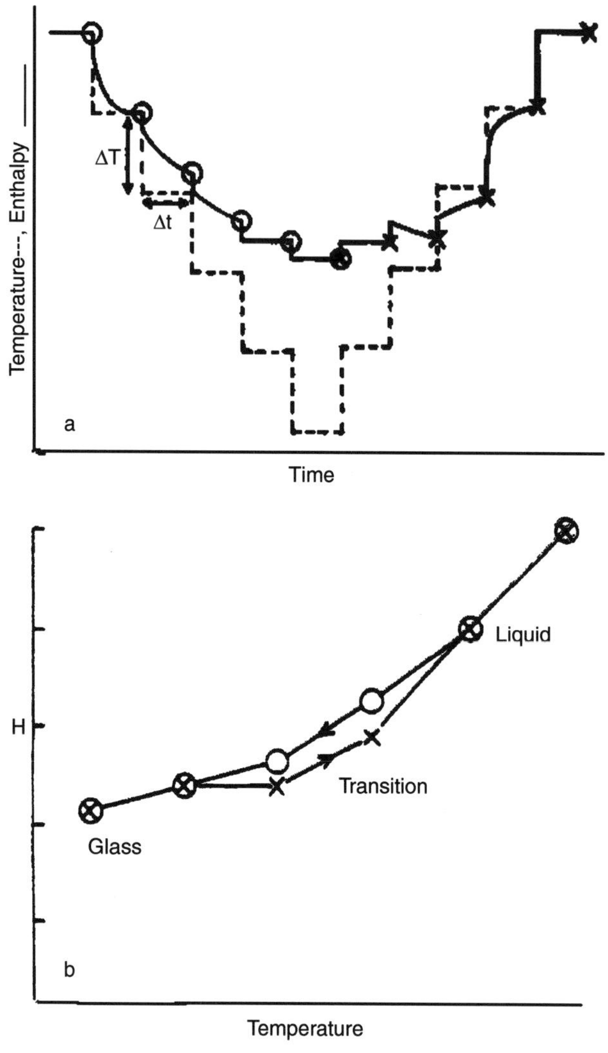

Figure 5 *Plots of (a) change in temperature, equilibrium enthalpy H_e and measured enthalpy H with time, and (b) change of H with temperature during stepwise cooling and reheating in the glass transition region. Reproduced with permission from Moynihan*[88]

The molecular origin of vitrification is still a subject of scientific debate, but clues as to the motions involved in the glass relaxation process are provided by the following observations:

(i) Nearly identical T_g values are obtained from specific heat and expansion coefficient measurements, suggesting that the readjustment mechanisms of energy and volume are closely similar.

(ii) The abruptness of the transition indicates that relaxation processes have very large temperature coefficients, *i.e.* the potential energy barrier that the relaxing unit needs to overcome must be considerably higher than those associated with the mean thermal energy (or equilibrium energy fluctuations).

(iii) Close relationships exist between dielectric relaxation and glass transformation, suggesting that relaxation processes, which lead to vitrification involve motions that are related to those performed by dipoles (and other molecules) when they jump from one quasi-equilibrium position to another one.

(iv) All liquids are believed to have (almost) the same viscosity at their glass transitions, usually taken as 10^{12}–10^{14} Pa s. This corresponds to a viscous flow rate of *ca.* μm year^{-1} or mm century^{-1}.

6.5 Amorphous States and Freezing Behaviour

Most materials destined for eventual freeze-drying are supplied in the form of unsaturated aqueous solutions; but the dried product is a solid. The drying process involves several phase transitions, *e.g.* liquid–solid and solid–gas, but water and solutes do not necessarily undergo the same phase transitions or at the same time. As discussed in previous chapters, when the temperature is lowered, solutions tend to undercool (*i.e.* remain liquid below the equilibrium freezing temperature) before freezing actually occurs. After the onset of freezing, and given moderate cooling rates, the solution will then remain in equilibrium with the solid phase (*i.e.* ice). The solutes remain in the concentrated, residual liquid phase. As the temperature is lowered further, the solution will become increasingly more concentrated. In principle, variations in the rate of cooling can affect the morphology of the product (*e.g.* ice crystal size distribution and specific surface area). In production-scale pharmaceutical freeze-drying, the rate of cooling (typically 0.2–0.3° min^{-1} or less) can always be considered as *low*, and minor variations hardly affect the dried product morphology. The situation is different for spray freezing, film freezing or hyperquenching, where cooling rates up to 10^4 deg s^{-1} can be achieved.

In multicomponent systems, solute primary crystallisation may or may not occur during freezing; it may also be incomplete, depending on formulation details. Solutions tend to supersaturation, *i.e.* water continues to freeze, but at a decreasing rate, and the solutes remain in a true solution of increasing concentration. At some characteristic temperature and composition, freezing stops, at least at a measurable rate. The resulting mixture (excluding the crystalline ice phase) has then reached its glass transition. This transition from a supersaturated, freeze-concentrated aqueous solution of very high viscosity to a brittle solid is operationally characterised by a specific value of the glass transition temperature. T'_g is therefore regarded as the temperature at which freezing (at a measurable rate) is complete. Knowledge of T'_g of a particular formulation is of the utmost importance for the design of a rational freeze-drying cycle because it is this temperature that should not be exceeded during the removal of ice by sublimation. Its value depends on the *composition* of the solid solution remaining at T'_g but is independent of the *concentration* of the initial aqueous solution. At T'_g, the solid solution phase always contains a certain amount of unfrozen water that can vary between 15% and 50% and can only be removed by its diffusion to the surface, followed by desorption and transfer to the condenser (secondary drying), see Chapter 9.

6.6 Materials Science of the Glass Transition: Its Relevance to Freeze-Drying

Although the general features of vitreous states and the mechanisms of their formation are identical for all materials, there are some aspects that are of unique importance to freeze-drying technology, especially as applied to pharmaceutical products.

CHAPTER 7

Excipients: Their Role in Rational Formulation Design

7.1 Definitions and Classifications

Pharmaceutical agents are rarely single-component preparations, if only because of the ubiquitous presence of residual water in all products. Two-component aqueous mixtures can also be ruled out in practice. A liquid solution that is to be freeze-dried to yield an acceptable solid dosage form will therefore contain several components, some of them carried over from downstream processing, *e.g.* chromatographic separations, while others are added prior to freeze-drying, for specific purposes, such as processing aids or for the stability enhancement of a dried product. The term "excipient", which does not occur in standard dictionaries, is now generally used indiscriminately to describe any substance(s) already present or added to the solution of the bioactive component prior to freezing. In a guide published by the International Pharmaceutical Excipients Council (IPEC), the term "excipient" is defined as:

> . . . substances, other than the active drug substance of a finished dosage form, which have been appropriately evaluated for safety and are included in a drug delivery system to either aid the processing of the drug delivery system during its manufacture; protect; support; enhance stability, bioavailability, or patient acceptability; assist in product identification or enhance any other attributes of the overall safety and effectiveness of the drug delivery system during storage or use.[89]

A less wordy definition appears in the *Oxford Concise Medical Reference Dictionary* as " . . . a substance that is combined with a drug to render it suitable for administration; for example in the form of pills. Excipients should have no pharmacological actions themselves". There are, however, indications that in certain situations, excipients do interact with the drug substance and can have an influence on its modes of deactivation.[90] Excipients are thus still regarded as an enigma: " . . . they do not treat a disease, but the drug industry cannot do without them".[91]

Apart from substances that have been carried over from prior processing stages, *e.g.* salts, the reasons for adding excipients are manifold, as are also their chemical origins. The latter can be classified as follows:

(i) Natural products, including polymers and derivatives.
(ii) Synthetic polymers and chemically modified polymers.
(iii) Natural products, modified with synthetic polymers.
(iv) Small molecules.
(v) Natural products, chemically modified with small molecules.

A short selection of less well-known excipients, their chemical groups and applications is given in Table 1. A more exhaustive compilation can be found in Apte and Ogwu.[89]

Many of the more recently reported excipient additives are subjects of patents or patent applications and have not been tested for general clinical use. However, the literature contains frequent references to "new" excipients. Usually they are not new in the chemical sense, but only in their claimed application field. For instance, erythritol, a well-studied polyhydroxy compound (PHC), has recently been reported as a "... new, multipurpose excipient".[92] As regards FDA-approved freeze-drying excipients, the list is depressingly short. It consists mainly of mono- and disaccharides, some amino acids and a few simple peptides. This is in a sense curious because some very efficient *in vivo* lyoprotectants, biosynthesised by many organisms to counter the stresses of drying either by freezing or drought, are diverse and manifold. We shall return to this subject later.

7.2 Attributes and Requirements

In pharmaceutical freeze-drying operations, excipients fulfil several functions. These include:

- pH buffering.
- Ensuring isotonicity of the reconstituted solution at point of use, *e.g.* saline solution.
- In-process chemical protection (lyoprotectants) of the bioactive component.
- Long-term chemical stability of the dried product.
- Improved mechanical strength (bulking) of the freeze-dried cake; prevent powdering.

Table 1 *Classification of selected pharmaceutical excipients*

Product	*Application*
Group 1	
Cyclodextrin	Solubilisers for hydrophobic molecules
Fibronectin	Protein stabiliser
Phosvitin	Protein stabiliser
Phytic acid	Protein stabiliser
Galacturonic acid	Contrast medium for MRI
Hyaluronic acid	Viscoelasticity
Spermidine	Inhibits lipid peroxidation
Group 2	
Hydroxypropyl–methacrylamide	Drug targeting
Polylactate–polyglutamate	Sustained drug delivery
Polyglutamic acid	Hydrophilic analogue of hydrophobic molecules
Polyfumaric acid	Enhances bioadhesion
Group 3	
Tocopherol polyethylene glycol (PEG) succinate	Antioxidant, solubiliser
Galactosylated poly(lysine)	Gene delivery vector
Polyoxyethylated castor oil derivative	Surfactant, emulsifier
Group 4	
Mannosyl glycerate	Enzyme stabiliser
N-methyl pyrrolidinone	Solvent
Sucrose laurate	Solubiliser
Group 5	
Chlorogenic acid chitosan	Confers water solubility at basic pH
Acylated polyamino acids	Oral protein delivery
N-octylglucoside	Assists proteoliposome preparation
Palmitoyl glycol chitosan	Controlled drug delivery

- Mitigation of deleterious effects of residual water.
- Achieving/improving economical processing times.

Commonly used excipients fall into several distinct chemical groupings, depending on the functions they are to fulfil during or after completion of the drying process. As regards pH buffering, isotonicity of the reconstituted solution is often the main requirement. In that case, phosphate-buffered saline solution (PBS) at the relevant concentration is a common excipient. On the other hand, if the protection of an active material, *e.g.* a protein, is necessary during the freeze-concentration stage, then the *optimum* buffer must be chosen for that particular process; common buffers include phosphate/citrate, bicarbonate and acetate. Tris, Hepes and other organic buffers also find application when

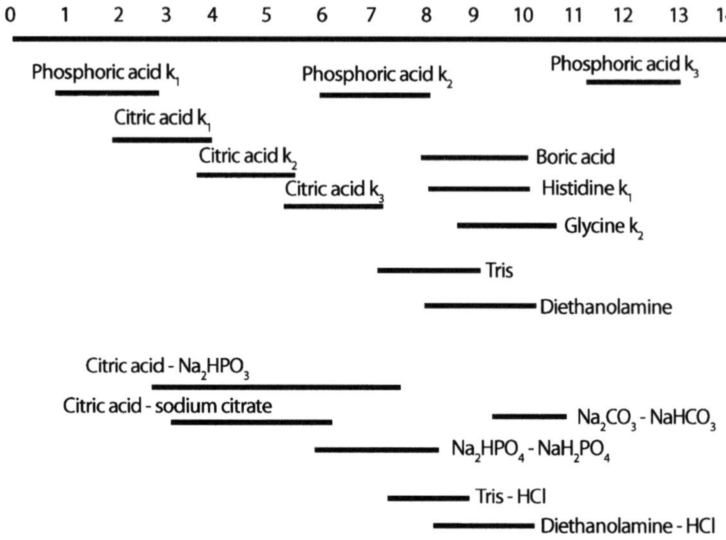

Figure 1 *Upper half: pK values of acids and bases ±1 pH unit. Lower half: useful pH ranges for some common buffering compositions. From Auffret*[93]

regulatory restrictions do not arise. Figure 1 shows pK values at 25°C of some common buffer acids, bases and binary mixtures.[93] Volatile buffers that would normally evaporate during the ice sublimation stage include ammonium acetate, formate and carbonate. Their use enables buffer-free dry products to be produced. Care must however be taken during reconstitution, because gases, especially CO_2, might become entrapped in the dry product and then give rise to irruptive effervescence when water is added to the product at the point of reconstitution.

Bulking agents are used simply to increase the total solids content in order to achieve a mechanically more robust plug. They find application especially when a low concentration of other solids has to be used, perhaps for reasons of isotonicity or low aqueous solubility of the active substance. Bulking agents, especially mannitol, also tend to improve the robustness and appearance of the plug, an important consideration in some markets. Any bulking agents used must, however, be able to crystallise *during* the freeze-drying process; only in this state will they have a neutral effect on product stability. By increasing the total solids content of the preparation, their use also acts unfavourably on the rate of ice sublimation.

More complex is the choice of excipients designed to act as lyoprotectants of chemically labile bioactive components. Earlier discussions have demonstrated that biopharmaceuticals, in particular, proteins, can rarely be induced to crystallise during freezing. They thus remain in the amorphous, supersaturated freeze-concentrate, where they become liable to inactivation by a variety of chemical processes, unless they are suitably protected by glass-forming excipients. Among classes of chemical compounds that find application as lyoprotectants, PHCs occupy pride of place, especially in pharmaceutical preparations. Peptides of various types are also used, although their glass-forming behaviour has not been investigated as intensively as has that of PHCs. Where data are available, they indicate glass transitions that are substantially lower than those of PHCs, possibly because of their zwitterionic nature.[94]

The essential physical requirements for an effective pharmaceutical lyoprotectant include:

- Chemical compatibility with other formulation components, both in solution and in the solid state.
- Adequate solubility in water and miscibility with other mixture components.
- Absence of inadvertent phase separation during freezing or thereafter.
- Absence of deleterious chemical reaction(s) involving excipients during freeze-concentration, *e.g.* glycation of peptides.
- "Suitable" glass transition temperatures for processing and dry storage (see later for details).
- Acceptable levels of *in vivo* clearance.

Stabilisers thus fulfil a dual role: they protect the labile drug component from damage during freezing by facilitating the formation of the glassy matrix; and by virtue of increasing the total solid content, they also provide physical strength to the dried plug. In many cases, the excipient (lyoprotectant) forms the major component of the formulated solution to be dried, so that its properties predominate in setting the glass attributes. The glass transition temperature is a function of the *total* chemical composition of the material, including water, but not ice. Any excipient present in an amorphous mixture will affect both T_g' of the solution and T_g of the dried product. Typically stabilisers are added to freeze-drying mixtures in order to chemically protect the active product against deterioration during the process. At the completion of the freezing stage, they must therefore be present in the amorphous solution phase. For successful freeze-drying, the solution must be cooled to yield a glass. As can be seen from Table 2, the choice of excipients will, to a large extent, determine the glass temperature, T_g', of the preparation, because the excipient is usually present as the major component. With careful selection, stabilisers will have a beneficial effect on T_g' and T_g, although some amorphous excipients can have a deleterious effect on product quality. The choice of dextran ($T_g' = -14°C$) as an excipient will permit drying at relatively high

Table 2 *Glass transition temperatures of selected anhydrous and aqueous, maximally freeze-concentrated excipients*

	T_g *(anhydrous) (°C)*	T_g' *(°C)*
Glycerol	−93	−100
Ribose	−10	−47
Xylose	−10	−47
Arabinose	−2	−43
Sorbitol	−3	
Mannitol	11 (?)	−28
Fructose	13	−42
Glucose	39	−43
Galactose	71	−42
Sucrose	70	−32
Maltose.H_2O	70	−29
Lactose	Cryst.	−28
α,α-trehalose (anhydrous)	106	−28
β,β-trehalose (anhydrous)	109	
Raffinose (anhydrous)	109	−26
Maltotriose	95	−24
Stachyose (anhydrous)	132	−24
Dextran	81†	−14
Byco A[a]	113	−14
Byco C[a]	116	−12
Byco O[a]	139	−15
Hydroxyethyl starch		−12[b]
Inulin	39[b]	
Hcit (anhydrous)	11	−53
Hcit.H_2O	−25	
Proline		−68
Glycine betaine		−98

Data from several sources.
[a] Non-gelling gelatin fractions, as commercially supplied (Croda Chemicals Ltd.), without additional drying.
[b] As commercially supplied, without additional drying.

temperatures, whereas the addition of glycerol ($T_g' = -100$°C) may lower the glass temperature to below the operating range of any condenser used in commercially available freeze-driers.

As regards the reliability of the entries in Table 2, reasonable agreement now exists between T_g' values reported in different publications. By contrast, T_g is often claimed to refer to a glass transition of the (implied) *anhydrous* substance. In practice, this presents problems mainly arising from residual water. The pre-1990 literature contains T_g data that suffer from more serious problems, arising from the water content of crystalline hydrates. Thus, trehalose is commercially supplied as a crystalline dihydrate. When melted and rapidly quenched, it does not recrystallise, but vitrifies. The reported T_g of the glass, 70°C, was for several years cited as the glass transition of trehalose, although it was actually the

Table 3 *Glass transitions of freeze-dried anhydrous citric acid and its sodium salts*

	T_g (°C)
Citric acid	11 ($T_g' = -53$°C)
Na citrate	69
Na_2 citrate	115
at pH 2.98	29
at pH 3.4	56
at pH 4.38	66

Data from Lu and Zografi[106] and Auffret (unpublished).

glass transition of trehalose.$2H_2O$, *i.e.* a glass that contained 66 mol% (9.5% w/w) of water.[74] This led to several misunderstandings until a solution of trehalose, rigorously freeze-dried, was found to have a T_g of 106°C. In order to determine T_g as accurately as possible, rigorous drying is therefore essential.

The combination of pH buffer with lyoprotectant functions can be utilised with those salts and acids that possess sufficiently high T_g values. Citric acid and its sodium salts present particularly interesting behaviour, a further illustration of the remaining mysteries of glasses. The results are summarised in Table 3.

7.3 Estimation of T_g of Complex Mixtures

Several methods, more or less rigorous, exist for the calculation of T_g from quantities accessible by calorimetric measurements. For binary (anhydrous) sugar systems, comprising glucose, fructose and sucrose, limiting values of (dT_g/dw_2) have been reported, based on the measured heat capacity differences between the glasses and the fluids at T_g.[78] It is interesting that the glass properties of sucrose do not resemble those of an equimolar mixture of glucose+fructose. A good working approximation for estimating glass transitions of binary systems is provided by the Gordon–Taylor equation, especially where one of the components is present in excess. It does not require knowledge of heat capacities but is based solely on the composition and the T_g values of the individual components:[95]

$$T_g = [w_1T_1 + kw_2T_2]/[w_1 + kw_2] \quad (1)$$

where T_1 and T_2 are the glass transitions of the two components and w_1, w_2 their mass (or mol) fractions; k is an adjustable parameter, the magnitude of which provides information about any specific interactions between the two components. Water is universally present, even in dried products, so that component 1 in Equation (1) usually refers to water. Although some uncertainty exists about the value, the glass transition temperature of water has traditionally been taken as 135 K.[96] Recently, however, and based on different methods of measurement, this value has been challenged, and 165 K has been suggested.[97] This value brings water more in line with other hydrogen-bonded network-forming systems. The value of the Gordon–Taylor equation, as shown, is limited, because water is invariably one of the components, and most pharmaceutical formulations therefore contain at least three components, including water. Equation (1) can, however, be expanded to make it applicable to any number of components,[78] although the procedure would require the knowledge of physical properties, including specific heat data of every component in the mixture; such information rarely exists. In practice, the application of the Gordon–Taylor equation can be extended, although in an approximate form to multicomponent mixtures; the procedure only requires data of their individual T_g.[98] Since the T_g of water is very low, as is also its molecular weight, even trace (<1% w/w) amounts of water will produce a marked effect on the glass transition of a mixture.

7.4 Effects of Excipients on Product Attributes

The influence exerted by excipients both on the process parameters and on the physical properties of the dried product needs to be clearly understood. Since the excipient is frequently the major component of a pharmaceutical mixture to be freeze-dried, it significantly increases the total solids content and hence the bulk density of the dried product, rendering it more stable against mechanical disruption. At the most basic level, a crystallising excipient will act in this manner without, however, affecting the *chemical* stability of any labile mixture components. An amorphous excipient, on the other hand, reduces the quantity of ice produced and thus the degree of freeze-concentration. In practice, this will shorten primary drying times but will, in turn, increase secondary drying times.

As regards the role played by excipients in determining the glass transitions of the freeze-concentrate and the final dried product, they may increase T_g', thus allowing the implementation of shorter cycles by raising the maximum safe temperature for ice sublimation. In addition, any excipient that increases T_g of the final dried product will thereby raise the product's maximum safe storage temperature. This is of particular importance for the production of shelf-stable products. Some excipients, particularly those of a PHC type, also stabilise proteins in solution against pH and/or salt denaturation,[99] while they also protect the protein, although by a completely different mechanism, against the damaging process of freeze-concentration.

7.5 Implications for Freeze-Drying

Depending on the physical behaviour and stability of the bioactive component, both amorphous and eutectic systems can be successfully freeze-dried but require different approaches. The most difficult cases are those that exhibit partial crystallisation, which may lead to uncontrollable intra-batch variability. Irrespective of the physical state of the system, maximum temperatures for ice sublimation (T_e or T'_g) can be identified. Above these temperatures, ice will melt at an appreciable rate, thus diluting the solution phase. The product may then exhibit mechanical collapse, and most probably it will also become subject to chemical changes (usually deterioration).

In a frozen amorphous system at T'_g, the product will often retain a significant amount of water, typically 20–50% w/w. At any stage thereafter, if the temperature is allowed to rise unduly, the residual glass softens into a viscous liquid with adverse consequences. Although most of the residual water must be removed, the amorphous matrix must be maintained in, or close to, the glassy state at all times. The maximum safe storage temperature after the complete removal of water will be given, at least for a limited period, by the glass temperature (T_g) of the final dried composition. For extended shelf lives, say months or years, glass transitions well in excess of the storage temperature must be aimed at. This aspect will be further discussed in Chapter 11.

In a eutectic system, the whole composition or just the excipient may be subject to crystallisation, but this process takes time to reach completion. It may be much longer than the time taken to freeze the product to the desired temperature (T'_g). After primary drying (ice sublimation), only solid "solutes" remain. The mixture may then be carefully warmed to its final storage temperature. The residual solid will not necessarily be anhydrous and may, for example, contain water of crystallisation. In addition, as the temperature is raised, the crystalline product may undergo solid–solid transitions, *i.e.* over a period of time, a different polymorph may become the preferred crystal habit. In a completely crystalline preparation, the maximum safe storage temperature will be governed by the component with the lowest melting point.

The physical state and the appearance of the dried cake are set during the freezing process, *i.e.* the kinetics of ice nucleation and crystal growth. Both can be controlled to some extent by variations in formulation and

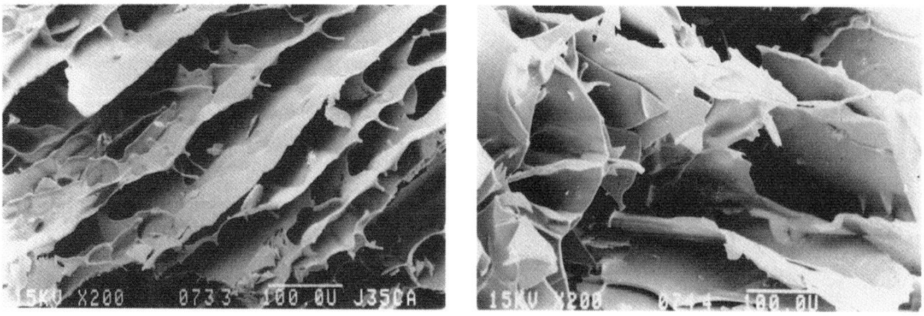

Figure 2 *Resulting cake structures after freeze-drying solutions of left: mannitol (crystalline) and right: sucrose (amorphous). Compare with Figure 9 in Chapter 5 for NaCl crystallisation from its solid solution in an amorphous sucrose matrix. Bar length = 100 μm. From Pafra Biopreservation, unpublished results*

process cycle (*e.g.* temperature). Some examples of different ice and product morphologies resulting from differences in the formulation and cooling rates are shown in Figure 2. It must be stressed, however, that both crystalline and amorphous states can yield acceptable products, albeit with different characteristics; some of them are summarised below:

Crystalline formulations

- Attractive visual appearance.
- Known safe storage temperature, governed by equilibrium melting points.
- Plug may become "powdery" if subjected to rough handling.
- Physical characterisation of a mixture can be complex.
- Crystallised excipients offer no chemical protection to labile biomolecules.
- Possibility of polymorphic and dehydration transitions during storage.

Amorphous formulations

- Cycle time can be shortened by choice of excipients with a high T_g'.
- Safe storage temperatures are governed by T_g that can be controlled.
- Hygroscopic, *i.e.* readily reconstituted into an aqueous solution.
- Easily characterised by thermal analytical methods.
- Glassy sheen sometimes found on plug surface is considered cosmetically undesirable.

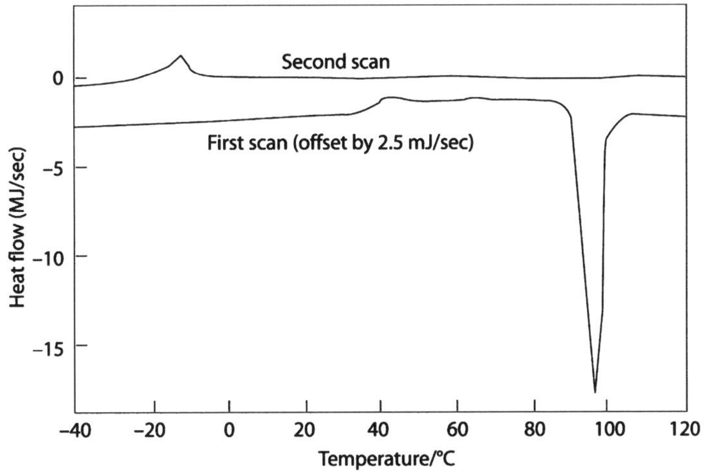

Figure 3 *DSC analysis of a freeze-dried preparation containing 3% protein, 95% lactose and 3% water. The first scan shows a T_g at ca. 30°C and a crystallisation exotherm at 90°C. The second scan shows an ice melting endotherm with an onset at –25°C. For details, see text*

7.6 Devitrification of Excipients in Dried Amorphous Preparations

In freeze-drying terminology, the term "devitrification" describes a process whereby one or several components of a solid amorphous product undergo crystallisation *in situ*. Where PHCs are used as stabilising excipients, their devitrification above T_g often leads to rapid bioinactivation of the product. Consider for example a freeze-dried preparation of a calcitonin gene-related protein (2%), stabilised with lactose (95%) and containing 3% residual water. Its behaviour on heating, as observed by differential scanning calorimetry (DSC), is illustrated by the lower heating scan in Figure 3. When the dried product was exposed to temperatures above T_g (313 K in this case), it became kinetically unstable and subject to unpredictable physical or chemical changes. In the mixture described here, the lactose crystallised irruptively at 360 K, leaving the residual amorphous phase, which now consisted of 40% protein and 60% "residual" water, made up of the original water content, less any water withdrawn from the solution by the crystallisation of lactose monohydrate. When the preparation was then cooled to 220 K and rewarmed (upper scan), an endotherm, symptomatic of an ice melting process, was observed. The protein had therefore been subjected to a second freeze-concentration process and had suffered complete inactivation, because it now lacked the previous protection afforded by a high concentration of carbohydrate excipient.

This type of sugar devitrification is therefore highly damaging and must be avoided. Usually this can be achieved by storage of the preparation below T_g, although isolated instances of crystallisation in glassy mixtures, stored well below T_g have also been reported.[74] By contrast to the damaging crystallisation of a PHC described earlier, the crystallisation of a *hydrate* from a previously amorphous preparation can also provide a beneficial desiccation effect, by removing water from the amorphous phase, thereby increasing its T_g and improving the storage stability. The process has been termed "self-stabilisation",[100] and its use is further discussed in Chapter 11 and illustrated in Figure 3 of Chapter 11. The degree of such desiccation depends on the mol ratio sugar:water and on the crystallisation rate. The stabilising effect is therefore expected to increase as the number of mols of water per mol of sugar increases, *i.e.* in the order

(α,α-trehalose, melibiose, melezitose).$2H_2O$ < mannotriose.$3H_2O$ < (β,β-trehalose, stachyose).$4H_2O$ < raffinose.$5H_2O$

7.7 Pharmaceutical Glasses: Formulation Aspects

During the development of a suitable formulation for a pharmaceutical, and especially a biopharmaceutical preparation, several important aspects need to be taken into account. They will not all be discussed in detail; hence, the following points should only serve as signposts, pointing the reader in the right direction.

Perhaps of overriding importance are pharmacokinetic aspects. They are particularly important for products with high protein contents. Thus, the chosen excipient(s) must on no account reduce or interfere with the functioning of therapeutic proteins. In other words, the chemical stresses that might be introduced into a solution during freeze-concentration must not lead to irreversible protein inactivation. The popular "water replacement" hypothesis to account for lyoprotectant effects of PHCs during drying ascribes the observed protection to the replacement, presumably reversibly, of so-called the hydration water molecules by sugar or alditol molecules.[101] On the other hand, the literature also contains references to irreversible "glycation" condensations between proteins and sugars, in which sugars are randomly (?) attached, presumably covalently, to sites on proteins, leading to their inactivation.[102,103] The extensively documented Maillard browning processes also bear witness to the deleterious reactions between sugars and peptides.[104] Since there is a long history of the successful use of PHCs as effective lyoprotectants, it would probably require extremes of pH to produce irreversible protein inactivation. Nevertheless, the effects of water removal during the freezing of a protein solution in excess PHC need to be investigated before a given PHC can be employed in further formulation development. Other properties of the bioactive protein that require knowledge include its specific activity under a variety of conditions and details of the denaturation/renaturation thermodynamics.

7.8 Constraints

Formulation development is beset by several constraints. Most important are probably the limitations imposed by regulatory requirements, mainly with regard to permission for the use of certain ingredients in pharmaceutical products. Restrictions imposed by regulators are particularly severe for parenteral preparations. Excipients currently in common use are included in Table 2. Several PHCs that have superior stabilising properties are not yet approved for pharmaceutical preparations because they cannot be prepared with a sufficiently high degree of purity or reproducibility. The polymeric condensation product obtained by the reaction of sucrose with ethylene chlorhydrin, and marketed under the trade name Ficoll, is one of the most effective stabilisers for freeze-dried proteins. Any excipients not currently approved would need to be registered which, in turn, would require much experimental evidence and the eventual registration of such compounds by the intending producer. The cost of carrying out the necessary work and of the novel excipient itself is often considered unacceptably high in relation to the benefit that such innovation may confer on the product in the marketplace. For useful discussions of regulatory requirements, as they currently exist in North America and Europe, the reader is referred to appropriate chapters in Cameron.[7]

Other constraints, more of a physical nature, which limit the choice of excipients, include solubility, stability, purity, the desired shelf life and appearance of the dried product, and special reconstitution requirements at the point of use. Thus, certain amino acids and oligopeptides, *e.g.* alanine and dialanine, provide good and chemically compatible stabilisers for protein-based products, but their use in practice is limited by their low solubilities. Where products are liable to exposure to high temperatures, perhaps during transport and/or storage, excipients with high T_g values are indicated. This excludes the use of amino acids and monosaccharides, even some disaccharides. Amino acids and peptides

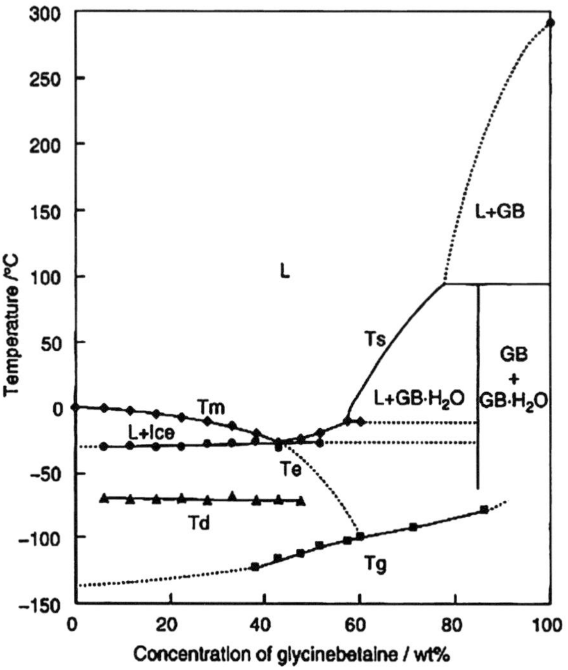

Figure 4 *State diagram of the glycinebetaine–water system. L: unsaturated solution; GB: crystalline glycinebetaine; T_s: solidus; T_d: devitrification temperature; dotted curves represent estimated portions of the diagram for which experimental data are still missing. Reproduced with permission from Komai and Murase*[94]

exhibit low glass transitions at physiological pH conditions; the values tend to be pH sensitive, presumably due to the zwitterionic character of such compounds. Some examples of amino acids and peptides are included in Table 2, and a recently published state diagram (incomplete) is shown in Figure 4 for glycine betaine.[94]

Because of their high glass transition temperatures, polymers can serve as useful excipients, especially when dried products require further processing at elevated temperatures or when products are exposed to high temperatures during shipping. As an example, scientists in the author's laboratory were able to prepare a dry phytase preparation, for incorporation at high temperature (superheated steam) into animal feed pellets. It was stabilised with Ficoll to a T_g of 145°C and a protein denaturation temperature of 160°C, while the enzyme in aqueous solution exhibited a denaturation temperature of *ca.* 65°C. The disadvantage of polymeric excipients can show up in long reconstitution times of dry products at the point of use. Their use is therefore not advised where rapid reconstitution is an absolute necessity, *e.g.* for products administered by injection in crisis cases.

7.9 Operational Degrees of Freedom

Based on these discussions, it follows that the scientist charged with developing a suitable formulation is often severely limited in the choice of excipients. Other things being equal, physical and chemical principles can, nevertheless, be applied and offer several lines of approach.

A choice exists between formulating a crystalline or an amorphous product. Although conventional wisdom in the pharmaceutical industry favours crystallinity, because of its believed association with stability, earlier discussions have pointed out the fallacies in such assumptions, especially where polymorphism or hydrate crystallisation cannot be ruled out. It is also found that freezing often produces metastable crystalline states that may eventually undergo transitions to more stable modifications. Thus, a recent report described a hitherto unknown monohydrate of mannitol formed during "non-aggressive" freeze-drying.[57] It was characterised by X-ray diffraction (see Figure 5), which produced a pattern with lines that could not be attributed to any of the known polymorphs. Differential thermogravimetry indicated a weight loss of 6% near 50°C corresponding to one mol water. The diffractogram, taken after removal of the water, contained only peaks that could

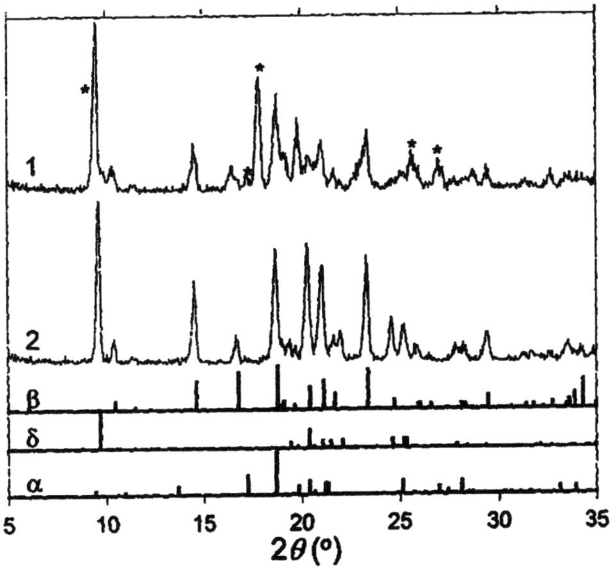

Figure 5 *X-ray powder diffraction patterns of (1) a freeze-dried mannitol solution and (2) the same preparation after heating at 70°C for 30 min. Peaks marked with asterisks do not appear in the diffractograms of any known mannitol polymorphs. All peaks in pattern (2) are attributable to β- or δ-mannitol. Reproduced with permission from Yu et al.*[57]

be assigned to the familiar β- and δ-polymorphs. The distinct probability must exist that there are other PHCs, with potential applications as excipients, which might exhibit a so-far undiscovered polymorphism.

Practical consequences of such metastable states and associated transitions for the freeze-drying of biopharmaceuticals are self-evident. Most damaging would be the uncontrolled release of water from a hydrate, giving rise to vial-to-vial variations in water content and water release during accelerated stability assays. Other consequences include the generation, but vial-to-vial variability, of polymorph mixtures.

Another degree of freedom relates to the choice of a suitable excipient/buffer system, quite distinct from ensuring the correct pH value. The complex phase relationships of water–PHC–salt systems have already been mentioned (see also Shalaev *et al.*[105]). It is imperative to be aware of possible effects of excipients on eutectic and glass formation in buffer mixtures.

The total solution composition and concentration, as well as the fill volume and vial type, may also be left to the product and process developers, at least within certain limits. This freedom of choice can facilitate the optimisation of the drying process. The planned residual water content of the final product may offer some degree of flexibility. It is determined by the drying conditions and, in turn, it sets limits on the safe storage temperature and the achievable shelf life of the product. Anecdotal evidence has been advanced for the dangers of overdrying proteins, although in the author's experience, almost complete drying (<0.1 g water per g solids), when expertly performed, does not lead to protein inactivation. It is likely, therefore, that damage arising from "overdrying" is actually the result of over-aggressive secondary drying, *i.e.* beyond the boundaries shown in Figure 5 in Chapter 9.

Finally, it is instructive to look to nature for a whole range of "new excipients" (actually very ancient!), which are synthesised *in vivo* by species that are exposed to the same type of water stress as that encountered during freeze-concentration. They are mainly amino acids, most of which do not occur in proteins, *e.g.* strombine [$(NH_2.COOH)_2$], the amino acid analogue of succinic acid, and ectoine, a cyclic condensation product formed by three lysine molecules. Indeed, much can be learned about useful formulation strategies from a study of natural water stress, and the strategies adopted by those organisms that can resist and/or tolerate seasonal drying, whether by freezing, drought or salination.[44] Despite their common *in vivo* occurrence and their undoubted protective action, it is unlikely whether the regulatory authorities would approve the use of such exotic excipients in pharmaceutical preparations, designed for mammals.

CHAPTER 8

Primary Drying: The Sublimation of Ice

8.1 Principles of Coupled Heat and Mass Transfer

In practice, at the beginning of the primary drying stage, more than 90% of the water in the initial solution has frozen. The ice is then removed by sublimation. Unlike the freezing process, ice sublimation is amenable to some measure of control. The heat flow to the ice front must be adjusted to balance exactly the heat absorbed by the sublimation of ice at the operating temperature of sublimation. In this chapter, we discuss the contributing mechanisms by which heat is transferred from the shelves of the freeze-drier to the ice front and the mechanisms by which water vapour is transferred to the condenser (mass transfer), and their relative contributions to the overall sublimation process.

8.2 Model

A typical pharmaceutical formulation will serve as model, frozen aqueous solutions that are contained in loosely stoppered glass vials. They stand directly on the freeze-drier shelves. Such a model appears deceptively simple but is in fact subject to considerable complexities that bedevil the modelling of optimum processing conditions and the estimation of a realistic drying cycle time. As emphasised in earlier chapters, for the case where the product does not display eutectic behaviour, it is imperative to freeze the solution to below T_g', in order to ensure maximal freeze-concentration and minimise deleterious chemical reactions that have been shown to occur at high rates in part frozen, supersaturated mixtures.[58,59] By the same token, the temperature during the removal of ice by sublimation must on no account be allowed to rise above T_g' at

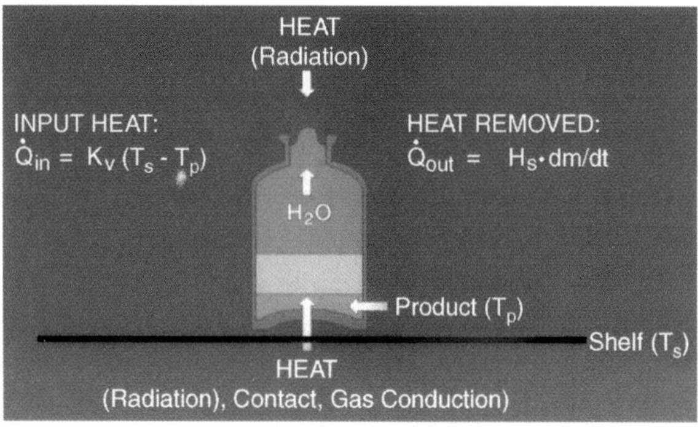

Figure 1 *The principle of coupled heat and mass transfer, where the energy supplied, Q_{in}, from the shelf balances the energy required to sublime water, Q_{out}, at a given mass transfer rate dm/dt. H_s is the latent heat of sublimation at the product temperature T_p*

any time during the sublimation process, so as to avoid ice melting back into the supersaturated mixture. With the setup shown in Figure 1, heat enters the product mainly, but not exclusively, through the bottom of the vial, whereas water vapour sublimes from the ice surface at the top of the vial. This results in a temperature gradient through the frozen plug, normal to the heating surface. There may also exist concentration (and density) gradients generated by the advancing ice front during freezing.

Depending partly on the chemical identities of the dissolved solids and partly on the way in which the freezing and any annealing processes have been performed, at the beginning of the drying cycle, the formulation may be composed of, in increasing order of complexity:

ice *and* crystalline solute(s)
or
ice *and* the freeze-concentrated, glassy solution
or
ice *and* (partially) crystallised solute(s) *and* a freeze-concentrated glassy component of unknown composition.

In practice, if cooling is performed correctly, most formulations will give rise to an amorphous glassy product during freeze-concentration. *Knowledge of the glass transition temperature T_g' is therefore absolutely essential.* It is frequently the lack of such knowledge that renders industrial freeze-drying processes as nothing better than trial-and-error exercises that are quite unlikely to yield the optimum process cycle. Starting from a known T_g', it becomes possible to estimate the optimum drying conditions to within reasonable uncertainty margins.

The view is sometimes stated, albeit erroneously, that the applied vacuum (*i.e.* the chamber pressure) provides the driving force for sublimation. The reader is reminded that washing, hung out to dry on a cold day, will freeze but will eventually dry, even without the need for a vacuum pump. The real driving force for sublimation, ΔG_s, is provided by the water vapour pressure difference between the ice front and the condenser surface. Since water vapour pressure is inversely proportional to the temperature, it is in fact the temperature difference between the ice front and the condenser surface that provides the driving force. This demonstrates that ΔG_s will be maximised by the choice of a formulation with a high T_g', combined with a condenser, which is able to operate at as low a temperature as practically possible. The chamber pressure affects the *rate* of sublimation, but not the driving force. For obvious reasons, the chamber pressure must be maintained below the saturation vapour pressure (SVP) of ice at the operating temperature.

8.3 Heat Transfer Mechanisms

The engineering principles and mechanisms of heat transfer that underlie the supply of thermal energy to the frozen product have been discussed in detail by Oetjen[7] and are only summarised here. Basically, the supply of heat energy is made up of three distinct contributing mechanisms:

(i) Radiation from "warm" surfaces ("warm" here refers to any temperature above that of the subliming ice front).
(ii) Conduction within and between solid phases (heating plates, glass vials, *etc.*).
(iii) Conduction within vapour phases produced by Brownian diffusion and molecular collisions (can also be expressed as convection).

Radiation effects are governed by Stefan's law. They become significant mainly near the walls and the door of the freeze-drier; they also depend on the material of the surfaces, usually coated or uncoated metal. For a radiating surface held at 30°C and an ice front at −30°C, the mass of ice sublimed by radiation alone has been estimated to be only of the order of 50 g $(h\ m^2)^{-1}$.[7]

Contributions arising from heat transfer within and between phases depend on the respective thermal conductivities of the materials, *e.g.* of the heating shelves, aluminium plates, glass vials, the chamber pressure, and the thermal conductivity of the frozen solution itself and of the portion of the plug that has already been dried. The proportions of the last two mentioned contributions change continuously during sublimation.

An approximate estimate of the effects of molecular collisions, *i.e.* convection, on the sublimation rate can be derived with the aid of the simple kinetic theory of gases. Thus, for sublimation from a free ice crystal surface:

$$\frac{dw}{dt} = \frac{K_V}{\Delta H}(T_c - T_s) = \frac{fp}{(2\pi RT_s)^{1/2}} \tag{1}$$

where

ΔH is the latent heat of ice sublimation;
w is the mass of ice sublimed in time t;
T_c is the temperature of the condenser;
T_s is the temperature at the subliming ice front;
K_V is the heat transfer coefficient;
p is the SVP of ice at T_S;
f is the "drying factor" to account for the probability of removal of water molecules before they can recondense; $0 \leq f \leq 1$.

For sublimation of ice, grown in an aqueous solution, Equation (1) may need to be substantially modified; for details see below.

In practice, the magnitudes of the various mechanisms to the total heat transfer, and thus also K_V, depend on the quality of the contact between solid phases, *e.g.* vial bottom and shelf or plate surface, and the chamber pressure, which is itself a function of the total number of gas molecules and the temperature. Figure 2 shows these relative magnitudes for three types of glass vials, standing on a polished stainless steel shelf, at two pressures, 14 and 52 Pa, respectively. The cross-hatched area represents heat transfer by conduction through and between solid phases (*e.g.* shelf, glass vial and the frozen plug), the shaded area represents the contribution from radiation, and the clear area is due to gas conduction (convection). An inspection of the two figures shows that in every case the major contribution to heat transfer is due to

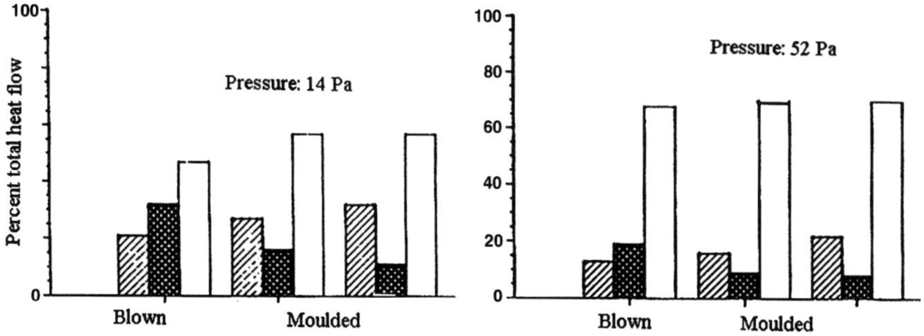

Figure 2 *Contributions of radiation (left), conduction (centre) and convection (right) to the total heat transfer for three vial types at two pressures. Redrawn with changes from Pikal*

collisions between gas molecules (convection), and this effect, in turn, becomes more pronounced with increasing chamber pressure. Conduction effects are seen to contribute a higher proportion of heat transfer for blown glass vials because their flat bottoms provide a better contact with the shelf than do moulded glass vials.

The relative contributions of the differing heat transfer routes therefore vary with pressure and with vial characteristics. A rise in chamber pressure may lower the product temperature by increasing mass transfer (sublimation) relative to heat input; this should be compensated by raising the shelf temperature or, preferably, by adjusting the primary drying time.

The total sublimation time t_s has been given as[107]

$$t_s = (\rho.w_1.\Delta H_s.w_1'.d)/\{(1/K_v) + (\kappa.d/2) + (\Delta H_s.\dot{w}.d/2)\} \qquad (2)$$

where

ρ = density of the frozen solution (kg/m^3);
w_1 = total water content (kg/kg);
w_1' = mass fraction of ice, frequently taken as 0.9;
d = fill depth (m);
κ = thermal conductivity of the frozen solution (kJ/°C.m.h);
K_v = heat transfer coefficient from the heat exchange fluid to the subliming ice surface;
$\dot{w}$ = water mass transfer (kg/h.m.Pa)

Equation (2) for t_s has been derived subject to two simplifying assumptions: (1) the dried cake does not offer any resistance to the transport of water vapour, *i.e.* it has a high porosity and that the frozen plug is non-porous; and (2) convective heat transfer within the dried portion of the plug can be neglected. The first assumption is at odds with some direct experimental results on the resistance of dried cakes to water vapour permeation.[108,109] They are shown in Figure 3 and discussed in more detail later. Altogether, these assumptions are said to introduce a *ca.* 2% uncertainty into the estimate of t_s.

Several experimental variables are required to enable Equation (2) to be applied. They include:

- the sum of heat transfer and thermal conductivity taken between the shelf and the subliming ice front;
- the thermal conductivity of the frozen plug;
- the mass transfer (sublimation) rate.

It is also useful to have some knowledge of the actual magnitude of K_v and how it varies with temperature, pressure and the configuration of

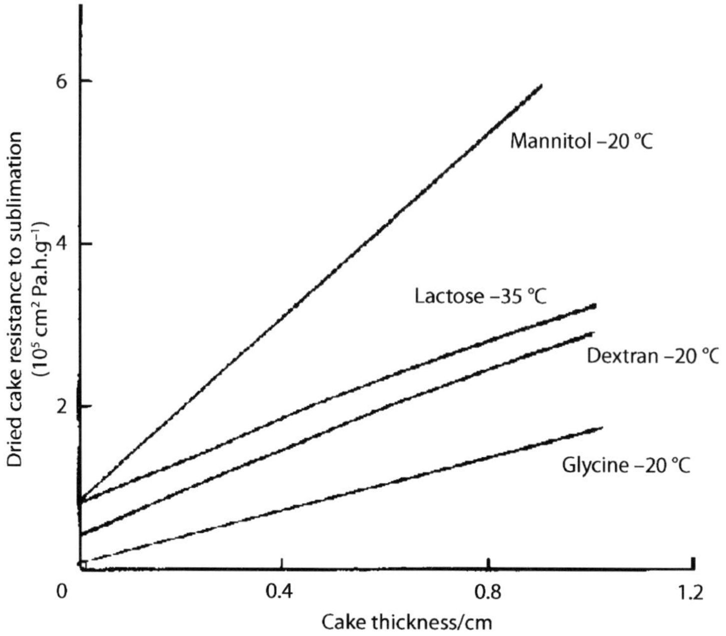

Figure 3 *Resistance to sublimation as function of cake thickness offered by different excipients at an initial solution concentration of 5% w/w. Adapted from Pikal*

the freeze-drier. This would then provide an estimate of ΔT, the temperature difference between the shelf and the subliming ice front.

Not many studies on pharmaceutical preparations have been performed in which all these parameters were carefully measured and documented. According to Oetjen, for a typical freeze-drying process in which a solution, containing 7% solids and contained in vials to a fill depth of 7 mm, is dried at −22°C, the calculated heat transfer coefficient K_v is 145 kJ (°C h m^2)$^{-1}$ and the mass transfer rate is then 1.3×10^{-4} kg (h m^2 mbar)$^{-1}$. Using this model system, it is instructive to evaluate and compare the relative magnitudes of the three terms in Equation (2); they are (from left to right) 2.33, 0.19 and 0.03, giving an estimated primary drying time of 2.5 h. Thus, the first term on the right-hand side is dominant and determines the drying time to an extent of *ca.* 90%. The fill depth is thus seen to exert the major influence on the drying time: t_s varies in a linear manner with d and inversely with ΔT. This result would of course argue against the approximation, discussed earlier, that the dried cake offers no resistance to ice sublimation. The thermal conductivity of the frozen solution appears to contribute little to the total heat transfer. Still using this model, a 3-fold increase of the fill depth would lead to a 4-fold increase of t_s to *ca.* 10 h. Experimental tests with different food products and a range of processing conditions have shown that Equation (2) is in most situations reasonably adequate for the prediction of t_s.

Figure 4 illustrates an experimental test performed on 5% solutions of KCl and polyvinyl pyrrolidone (PVP), both at 10-mm fill depth. The inverse relationship between t_s and T is clearly shown and, interestingly, takes a very similar course for both products, despite their marked physical differences, *i.e.* KCl is crystalline and PVP is amorphous. To a first approximation, the sublimation rate increases 3-fold for every 10° rise in temperature. It must, however, be remembered that for a wholly or partly amorphous product, the temperature of the subliming ice front must not be allowed to exceed the glass transition temperature T'_g. It follows that the product composition also plays an important role in determining t_s.

The effect of chamber pressure on ice sublimation also needs to be considered. It is shown for a particular case in Figure 5. Since gas conduction (molecular collisions) accounts for the major contribution to mass transfer (see Figure 2), an increase in pressure is expected to accelerate sublimation. The upper limit is set by the SVP of ice, but there is no advantage to be gained from evacuating the chamber to a pressure substantially below the SVP.

The heat transfer coefficient K_v increases with chamber pressure typically for pressures in excess of 10 Pa. The effect of chamber pressure on the sublimation rate can be estimated. It is necessary, however, once again

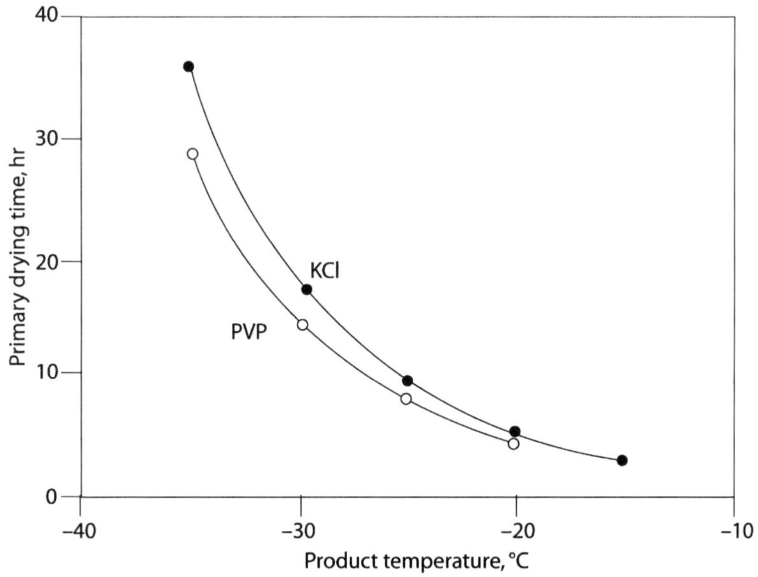

Figure 4 *Primary drying time dependence on the product temperature for a crystalline (KCl) and an amorphous (PVP) product. Adapted from Pikal*

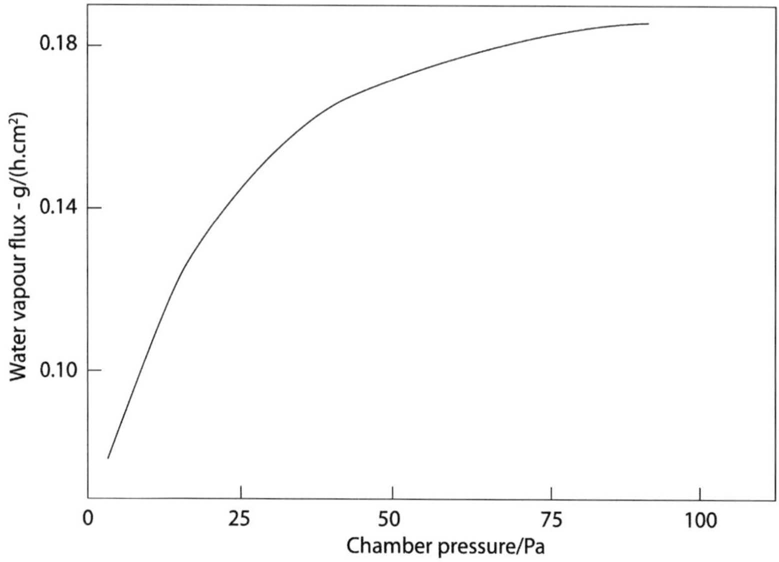

Figure 5 *Effect of chamber pressure on sublimation rate. Fill volume: 2 ml; vial diameter: 2.54 cm^2. Adapted from Pikal*

to take into account the various mechanisms by which heat is transferred between the shelf surface and the subliming ice front.

It is good practice to adjust the chamber pressure (but keeping below the SVP of water vapour over ice at the drying temperature) by a dry, non-condensing gas bleed (nitrogen, argon and helium). As a generalisation the pressure might be set close to, but not exceeding, the SVP of water over ice at the product temperature. A typical relationship between the sublimation rate and chamber pressure is shown in Figure 5. If the pressure is allowed to fall too far, then the rates of heat transfer into the product will no longer "balance" the rate of mass transfer and the product temperature will decrease (sublimative cooling). For reasons of process efficiency, the chamber pressure should probably not be allowed to drop to below 50% of the SVP at the relevant temperature.

8.4 Influences of Formulation and Freezing Procedure on Ice Sublimation

There are additional factors to be taken into account that influence t_s, and also the quality and uniformity of products freeze-dried in vials:

- ice crystal size distribution;
- solid content (solution concentration);

- concentration gradients (concentration polarisation, see Chapter 4);
- formation of an amorphous, glassy product "skin".

Of these, the ice crystal size distribution is set early on in the process during freezing by the nucleation and growth conditions of the ice crystals, which, in principle, depend on the applied cooling rate. A qualitative summary of the effects of cooling rate on ice crystal size distribution is given in Table 1. Definitions of "fast" and "slow", as applied to ice crystallisation, deserve a mention; they have already been touched upon in Chapter 4. Since the linear growth rate of ice in pure water and in dilute solutions is very high, then in industrial practice, actual cooling rates must always be considered as low. High rates can be achieved for small samples with the aid of cryogens by methods such as spin freezing or drop freezing.

The details of ice growth set the texture and porosity of the final, dried product. To illustrate this point, some examples of freeze-induced textures are shown in Figure 2 in Chapter 7. Any thermal treatment of the frozen solution, *e.g.* annealing or thermal cycling, can only lead to an irreversible coarsening of the texture, referred to as "maturation". The relative advantages and disadvantages of targeted maturation have been discussed at length, especially in the food processing literature. An example of maturation of ice crystals in frozen meat is shown in Figure 15 in Chapter 4. From the point of view of ice sublimation, a coarsened ice crystal structure reduces t_s, but it may also produce undesirable side effects.

The other factors, mentioned earlier, relate to details of the particular formulation. Thus, the effects of a high solid content, when coupled with small ice crystals, increase the likelihood of a complete coating of ice crystals by the glassy product matrix. Such a film coating would markedly reduce the ice sublimation rate, as given by Equation (1), where a "retardation factor" would need to be introduced to allow for the water vapour diffusion across the amorphous layer. High solid contents, especially of polymers, would also give rise to solute concentration polarisation ahead of the advancing freezing front (see Figure 14 in Chapter 4) and, in extreme cases, would produce an almost impermeable

Table 1 *Effect of cooling rate on ice crystal dimensions and primary drying time*

Cooling rate	*Nucleation rate*	*Number of ice crystals*	*Size of ice crystals*	*Ice sublimation time*
Low	Low	Small	Large	Shorter
High	High	Large	Small	Longer

glassy product skin at the surface of the frozen solution. It is seen, therefore, that formulation details, as well as cooling rates, can markedly affect the subsequent ice sublimation process. It has been suggested[110] that the actual ice sublimation rate, as expressed in an idealised form by Equation (1), may well be lower in practice by factors in excess of three orders of magnitude for small ice crystals that are completely coated by glassy films of freeze-concentrated solution. In summary, therefore, rapid cooling results in small ice crystals (a large specific ice surface area) and provides a uniform texture in the dried product, but because of lack of ice connectivity, such crystals are also likely to reduce the sublimation rate to an economically unacceptable degree. Coarse structures are usually preferred because they favour ice sublimation, the downside being that, depending on the solid content, product "skins" may develop at the surface of the advancing ice front.

The actual drying (sublimation) time t_s is thus seen to depend on the following quantities:

Fill depth (m)
Density of the frozen product (kg m^{-3})
Water content of product solution (kg kg^{-1})
Latent heat of sublimation at the particular drying temperature (kJ kg^{-1})
Temperature difference between shelf and ice front (°C)
Total heat transfer coefficient from shelf to sublimation front [kJ (m^2 h deg)$^{-1}$]
Thermal conductivity of frozen product [kJ (m h deg)$^{-1}$]
Fraction of *frozen* water (usually > 0.9)
Permeability (*i.e.* porosity) of dried product to water vapour [kg (m h Pa)$^{-1}$]

8.5 Interplay of Pressure, Temperature and Time

The complex relationship between chamber pressure and temperatures of shelf and product, and its impact on the sublimation rate, is illustrated in Figure 6 for a recombinant protein preparation.[111] Let us assume that T_g' for such a composition is –20°, so that the drying conditions must be set to ensure that the product temperature does not exceed this value at any time during the primary drying cycle. The sublimation rate corresponding to point A, at a chamber pressure of 40

Pa, is seen to be 0.16 g cm^{-2} h^{-1} at −20°C. If the pressure is halved, without a change in the shelf temperature, the product temperature falls to −25°C (B) and the process will therefore be accompanied by a marked decrease in the sublimation rate. If, on the other hand, the pressure is raised to 55 Pa, still at a constant shelf temperature, then the sublimation rate increases to 0.19 g cm^{-2} h^{-1}, because the product temperature rises to −17.5°C, which is, however, well above T_g' and must therefore be avoided. By changing the pressure, as well as the shelf temperature, represented by point D, the sublimation rate can be increased without danger of overheating the product.

It must be emphasised that the actual numerical values used in the construction of Figure 6 are hypothetical and refer to a particular set of conditions for product composition, vial type, fill depth, model of freeze-drier and degree of shelf loading. However, similar diagrams could be constructed for any freeze-drier/vial/loading combination. Furthermore, Figure 6 takes no account of discontinuities in the thermophysical properties of the product at T_g'. This discussion nevertheless serves to illustrate the general principle of interplay between shelf

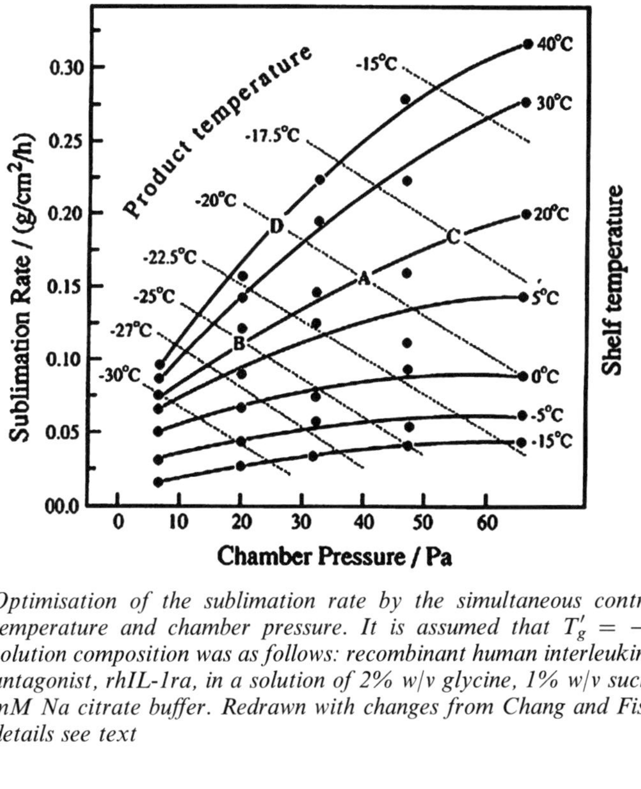

Figure 6 *Optimisation of the sublimation rate by the simultaneous control of shelf temperature and chamber pressure. It is assumed that* $T_g' = -20°C$. *The solution composition was as follows: recombinant human interleukin-1 receptor antagonist, rhIL-1ra, in a solution of 2% w/v glycine, 1% w/v sucrose and 10 mM Na citrate buffer. Redrawn with changes from Chang and Fisher.*[111] *For details see text*

temperature and chamber pressure in determining the sublimation rate at a constant product temperature ($<T_g'$).

8.6 Adjustable Parameters: Formulation, Concentration and Fill Depth

Setting the product composition of a pharmaceutical preparation, so as to effect an efficient and economical drying cycle, will probably always be beyond the remit of the process engineer. However, bearing in mind the complexities of the freeze-drying process, its optimisation should be given due consideration even by those responsible for product development.

The influence of the solid content makes itself felt during the later stages of the drying process. At the beginning of the drying cycle, heat is supplied to the shelves in order to balance the latent heat of sublimation absorbed by the ice. As the ice front migrates downwards during sublimation, the network of the dried product offers an increasingly strong resistance to further sublimation; the drying rate therefore decreases. The effect of the dry product resistance is difficult to estimate accurately because it depends very much on the actual texture (porosity), which, in turn, is determined by the primary freezing process and any subsequent annealing accompanied by maturation. It was shown earlier how ice nucleation and crystal growth kinetics depend in part on the solid content of the solution. It follows that a high solid content will decrease the porosity of the dried portion of the plug and hence raise the resistance to further sublimation.

On the other hand, an excessive reduction of the solid content results in a plug that is brittle and mechanically too weak to withstand rough handling of the containing vial during packaging and shipping. Although neither the chemical stability nor the efficacy of the dried product will necessarily suffer from a broken-up plug, its appearance is held to be undesirable, even unacceptable, in some markets.

The important role played by the solution fill depth in determining the sublimation rate has already been touched upon and can hardly be overstated. It depends on the shape of the containing vessel in relation to the fill volume. Ideally, a fill depth of 5 mm is to be recommended. Freeze-drying of chemically labile products at fill depths in excess of 20 mm is to be avoided. Where for pharmacokinetic or other reasons (*e.g.* solubility) a certain volume is prescribed for the administration of the reconstituted solution, the limitation of fill depth has to be achieved by other means. This often involves the use of a vial with a larger-than-ideal diameter. For reasons of cost and suboptimal utilisation of freeze-drier shelf space, such practice may seem to be uneconomical, but as regards product quality and drying time, it is to be preferred over excessive fill depths.

The effect of fill depth can best be demonstrated by considering the resistance of the dried cake to further sublimation. Figure 3 illustrates the effect of cake thickness (fill volume) on the resistance to sublimation for a series of products with the same solid content, 5% w/w. The linear relationship between fill depth and resistance suggests the absence of a concentration gradient within the frozen solution, *i.e.* an indicator of good freeze-drying practice. The reason for the markedly greater slope for mannitol has not been explained; it might be due to the possibility that mannitol had crystallised during the process. None of the lines passes through the origin, indicating that not too much significance can be attached to the actual numerical values. Nevertheless, the relationship between fill depth and plug resistance to sublimation is demonstrated.

8.7 Ideal Conditions

A set of parameters now emerges, the correct combination of which would produce ideal conditions, at least from the point of view of freeze-drying. This combination would offer an efficient drying cycle producing a high yield of dried product with an acceptable storage stability. Unfortunately, the frequently used device to maintain a high sublimation rate is just to heat the shelf, but this practice is to be discouraged. It may well cause the product temperature at the bottom of a vial to rise to above T_g'. Preferably the shelf temperature should be set to its maximum calculated value at the beginning of the drying cycle, and to be reduced gradually, to arrive back at T_g' on completion of ice removal. Such a protocol may however be difficult to implement in practice because the site of cooling, the ice front, is remote from the site of heating, *i.e.* the vial base.

It is thus established that moderate concentrations and fill depths facilitate ice sublimation. To arrive at a reasonable compromise, a solid content in the region of 10% w/w should be aimed at. Coupled with a fill depth of *ca.* 5 mm, this provides a good compromise to achieve an efficient primary drying protocol for many pharmaceutical formulations.

8.8 Accelerated Freeze-Drying

In efforts to speed up ice sublimation, the use of a variety of additives has been proposed. For example, tertiary butyl alcohol (TBA) has been claimed to accelerate the primary drying rate, partly because of its high vapour pressure.[112] In ternary model mixtures of water–sucrose–TBA, no specific interactions between sucrose and TBA were noted, implying that TBA does not affect the T_g' value (241 K), usually ascribed to aqueous solutions of sucrose. In other words, TBA is excluded from the amorphous freeze-concentrate and crystallises completely from the frozen mixture. Because of its high vapour pressure, it is presumably sublimed along with the ice without leaving a residue in the dried, amorphous cake. The specific effect of TBA on the system has been claimed to be solely related to a modification of the ice crystal habit, leading to the growth of needle-shaped crystals. This produces a significant increase in the specific surface area available for sublimation and a concomitant decrease in the resistance of the dried cake to mass transfer. The high vapour pressure of TBA provides additional sublimative cooling, "allowing" shelf temperatures to be set well above T_g'. Consideration should, however, be given to the stability of the product as the rate of TBA and ice sublimation decrease.

Claims have been made that the TBA does not enter into the product glassy phase.[112] It has, however, also been reported that some TBA is indeed incorporated into the freeze-concentrate.[113] If true, this might raise regulatory problems with pharmaceutical product registration. It is also not yet clear whether at temperatures in the region of T_g', the effect of TBA is unique. For instance, with sucrose–water systems, the presence of a volatile buffer salt (ammonium bicarbonate, acetate, formate) has a more marked effect on the sublimation rate than has TBA. On the other hand, with PVP–water systems, TBA is superior in enhancing the sublimation rate.[113] It thus appears that the choice of excipient can also affect the drying rate.

There are, however, other ways in *which* conventional drying could, in principle, be performed at temperatures above T_g', without deleterious results. If a solution could be treated so that one or several components can undergo crystallisation, then the crystals formed would serve as a substitute for the subliming ice crystals and could mechanically protect the porous, amorphous matrix against structural collapse. The efficacy of this type of formulation for freeze-drying purposes has been demonstrated for the system water–sucrose–NaCl.[114] Its practical applications may, however, be limited to products where the bioactive component is not chemically degraded in the freeze-concentrate at temperatures above T_g'.

Finally, and for the sake of completeness, mention is made here of freeze-drying from organic solvents above ambient temperatures that, in principle, offers attractive possibilities. Solvents are used where aqueous miscibility of the drug substance is very limited and presents problems with formulation. Organic solvents can also provide a stable environment for water-sensitive compounds and, like ice, they can be removed by sublimation to yield dry cakes that can be readily reconstituted. The best strategy is to use a solvent mixture with a convenient eutectic point, for instance dimethylsulfone + 1,1,1-trichlorobutanol, having melting temperatures of 109 and 78°C, respectively. The mol ratio of the eutectic mixture is 40:60, with a T_e of 50°C. Primary drying of a formulated progesterone can be performed with a starting temperature of 40 °C, ramped at regular intervals up to 60°C. The drying cycle requires only *ca.* 7 h, with a similar period for secondary drying (see Chapter 9). Reconstitution for most products can be performed in an aqueous/organic solvent mixture.

Provided there are no regulatory problems or fire/explosive hazards, the advantages of this type of drying can be summarised as follows:

- Sparingly water-soluble substances can be treated, provided there are no indications of toxic effects.
- Lyophilisation takes place without the need for refrigeration of the chamber (a refrigerated condenser is of course required).
- Short drying cycles depend on the vapour pressure of the solvent(s).
- Acceptable cake structures with minimal residual solvent contents and fast reconstitution in mixed aqueous solvents can be achieved.

8.9 Monitoring Ice Sublimation

It has already been mentioned that there exists no absolute dividing line between primary and secondary drying. However, for practical purposes of process control and economy, it is important to establish the point of completion of sublimation, *i.e.* the removal of *all* ice from the product. The interested reader is directed to the literature, where available techniques have been described and compared in detail.[115] The most direct indicator of ice removal is the chart recorder output of the product temperature. As sublimation nears completion, the product temperature rises, finally to reach the shelf temperature. A practical problem lies in the difficulty of measuring the temperature of the dried cake in a reliable manner, and in the atypical drying behaviour of any vial that carries a temperature probe.

The "pressure rise" test, although disliked by some practitioners, is a popular method for determining the endpoint of sublimation. By shutting off any gas flow into the chamber and isolating the condenser and the pumping system from the chamber, the chamber pressure is monitored. The absence of a pressure rise, within a period of *ca.* 2 s, is taken as an indication of the complete removal of ice. The necessary period of monitoring will depend on the time taken to fill the chamber with water vapour, and therefore on the total number of vials in the drier.

Other more refined methods that rely on pressure difference include the combination of different types of pressure probes, *e.g.* a combination of a capacitance pressure probe and a Pirani gauge.[115] The former device measures the total pressure, whereas the Pirani reading is affected by the thermal conductivity of the gas, *e.g.* water vapour. Completion of sublimation is indicated when both pressure readings coincide.

Anemometers (wind mills) have been described as particularly sensitive methods of detecting gas flow in freeze-drier chambers. Such devices cannot, of course, discriminate between different gases.

The incorporation of mass spectrometers in freeze-driers, even in production equipment, is a recent development. Their use offers a particularly sensitive method for monitoring the completion of sublimation. As an added advantage, they will also detect the onset of any product deterioration and/or decomposition.

8.10 Summary

In summary, the practical considerations and conditions for effective ice sublimation are:

- temperature difference between product ice front and condenser;
- chamber pressure;
- structural continuity of the ice phase throughout the frozen solution;
- volume of water to be removed;
- total solid content and chemical composition of the product;
- liquid fill depth (*i.e.* container geometry);
- design and performance efficiency of the freeze-drier.

For reasons of economy, short drying cycles are of course highly desirable, and much time and effort are often devoted by the pharmaceutical industry to improve (shorten) historical drying cycles. A mature consideration of these seven factors must, however, lead to the important conclusion that for any given product formulation and vial type, there is a limit to any process cycle reduction that can be achieved, without putting the quality of the dried product at risk.

CHAPTER 9

Secondary Drying: The Removal of Unfrozen Water

9.1 Unfrozen (Residual) Water

After a multicomponent aqueous solution has been freeze-concentrated to the limit and the ice has been sublimed, any residual unfrozen water must be removed from the remaining solid solution by diffusion, desorption and evaporation (transfer to the condenser). This process is termed "secondary drying". For an amorphous preparation, the amount of unfrozen water remaining after the removal of ice may be typically 20–30% w/w, but much higher values, even >50% w/w, have been found in some formulations. Attempts are on record to measure the amount of water that remains in the freeze-concentrated solution phase at T'_g, mainly by differential scanning calorimetry (DSC),[116,117] but such measurements are subject to considerable experimental uncertainties, which will be further discussed in Chapter 11.

Much lower amounts of water remain in crystalline systems, where water only exists in the form of a thin adsorbed layer on the crystal surfaces. Crystalline hydrates form an exception; here the water molecules are hydrogen bonded to one another and/or to the solute molecules, usually in stoichiometric ratios. They are thus an integral part of the crystalline compound. Important examples from the point of view of freeze-drying are $NaCl.2H_2O$, mannitol.H_2O, α,α-trehalose.$2H_2O$[†] and raffinose.$5H_2O$. In the first case, the hydrate contains 38% w/w of water. If, therefore, the dihydrate is able to crystallise during the freezing

[†] It appears to be generally assumed that the word "trehalose" used in the literature refers to α,α-trehalose; hence the mention of the dihydrate. However, its less well publicised isomer β,β-trehalose crystallises as a tetrahydrate. This is a major difference that has important implications for the behaviour of trehalose as excipient (C.J. Roberts and F. Franks, *J. Chem. Soc. Faraday Trans.* 1996, **92**, 1337–1343).

process, this amount of water will not be removed completely by sublimation, at least not during a normal primary drying process. In the case of mannitol, the hydrate contains only 9% w/w of water. On the other hand, a typical formulation is likely to contain substantially more mannitol than NaCl, so that damage caused during decomposition of the crystalline hydrate, with the concomitant liberation of water, would cause more damage to the product.

In general, no distinct boundary exists between primary and secondary drying because the removal of unfrozen water from the amorphous matrix above the receding ice front may commence even while ice is still subliming from lower regions of the frozen plug. Compared to ice sublimation, the secondary drying process is poorly understood and less accessible to quantitative analytical treatment. Thus, a safe required drying period can only be estimated. The most important point is that, even after ice sublimation is complete, an amorphous product must be maintained at temperatures close to the glassy state, T_g'. This is in stark contrast to advice sometimes found, especially in the technical literature, that immediately heating the material to room temperature does not produce adverse effects.[115] If, during the drying process, T_g is exceeded for any length of time, the fragile structure of the drying cake will begin to collapse because the viscosity of the solid matrix decreases rapidly with increases in temperature above T_g. This phenomenon is illustrated in more detail in Section 7. Figure 1 illustrates the onset of such collapse in a solution during the initial stage of secondary drying, even when carefully performed. Apart from progressive structural collapse, deleterious processes,

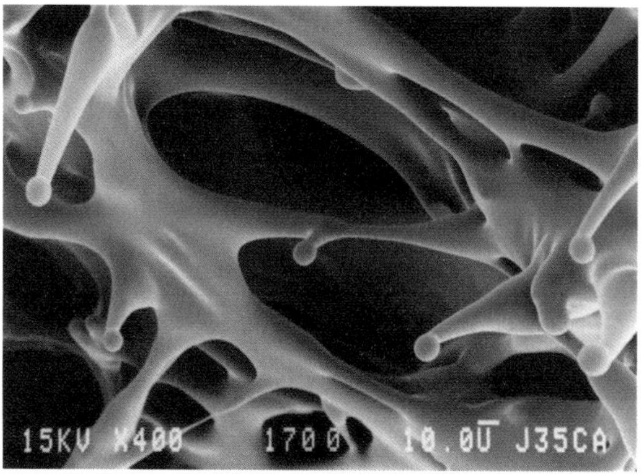

Figure 1 *Scanning electron micrograph of freeze-dried sucrose; the scale bar corresponds to 10 μm. Globules, formed at the ends of the filaments, are evidence of minor collapse during drying, when the temperature exceeded the softening point for short periods*

such as protein aggregation, previously retarded by the very low diffusion rates in the glass, are then also likely to proceed, often at rates substantially higher than in dilute solution at ambient temperatures. This is but one example of how product quality can be sacrificed in efforts to improve process economy.

9.2 Removal of Unfrozen Water by Diffusion

Although the process of unfrozen water removal from the material, subsequent to ice sublimation, is commonly referred to as "desorption", this is misleading. It has been shown that the residual water forms a mobile component of a solid solution, so that "diffusion" more correctly describes the mechanism of its removal. Compared to the diffusion rate of water from the bulk to the surface, its eventual desorption from the surface is rapid and can in practice almost be neglected. As will be shown later, secondary drying kinetics can be modelled adequately by standard treatments of diffusive processes. The effects of many variables on the kinetics of secondary drying may be found in Pikal *et al.*[118]

9.3 Effects of Pressure

For the reason mentioned earlier, and in contrast to primary drying, chamber pressure (a controllable variable) does not appear to influence secondary drying rates to a significant extent. It is also known that neither mannitol (crystalline) nor a formulated amorphous moxalactam product exhibit systematic effects of pressure on drying kinetics. However, although for a given product the drying rate is independent of pressure, different products at the same pressure will experience differing secondary drying rates. The almost universal practice of conducting secondary drying at very low pressures, does not, as is popularly supposed, increase the rate of water removal.

9.4 Effect of Water Content

At the completion of primary drying, the water content of an amorphous product is sometimes expressed as W'_g g water per g solid. The actual value, typically 0.3–0.5 g g^{-1}, is governed by details of the freezing process and will vary from product to product, or even from vial to vial. For most practical purposes, however, it may be assumed that the water content has little or no effect on secondary drying. This

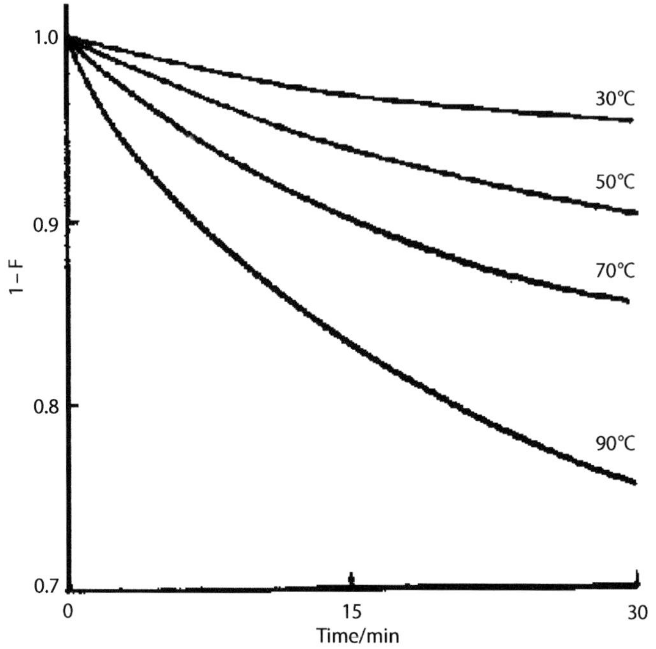

Figure 2 *Isothermal drying profiles of Ficoll glass film of 0.08-mm thickness, where F is defined as the fractional extent of drying, the mass of water removed at time t, divided by the initial water content. Thus $F = 1$ corresponds to the initial water content, 8.87% in this case. Redrawn, with alterations, from ref. 119*

has been established for freeze-dried preparations and also for the evaporative drying of polymer films with different initial water contents, as shown in Figure 2.[119] Thus, the independence of drying rate from water content is a general property of plasticised glassy materials and is not limited to disperse (freeze-dried) systems. The non-linearity of the drying isotherms indicates that diffusion within the sample, and not evaporation from the surface is the rate-limiting step. The analyses of the isotherms also demonstrate that water is mobile in the glass, even well below the measured glass transition.

9.5 Effects of Fill Depth and Cake Morphology

The *specific* surface area of the plug is generally quite large, typically up to 3 $m^2\ g^{-1}$, and will not only vary between different formulations but also with processing history. For example, the specific surface area of the ice crystals normally increases as does the cooling rate and, hence the degree of undercooling. Thus, increasing the fill volume that will effectively lower the cooling rate and therefore also the degree of undercooling would be expected to decrease the specific surface area of a freeze-dried product. It has also been found that, the specific surface area decreases as the concentration of the feedstock solution increases.

Pikal *et al.*[118] studied a 5%-povidone (polyvinyl pyrrolidone, PVP) solution, dried from an 8-ml fill volume, of 2-cm depth, and from a 4-ml fill volume at a 1-cm fill depth. In this instance, the fill depth had little effect on the specific surface area of the dried product (2.5 $m^2\ g^{-1}$ compared to 2.3 $m^2\ g^{-1}$). It was found that the rates of secondary drying (as measured by the mean ratio of the $(1 - F)$ values), the normalised water contents of the two products were essentially identical. A mean ratio of $(1 - F)$ of 1.19 ± 0.17 was found, indicating that cake thickness did not significantly affect secondary drying kinetics. It can thus be concluded that material at the top of the cake dries at a similar rate to material at the bottom, provided overheating is avoided. Similar experiments were carried out on an amorphous, formulated moxalactam product, as shown below.

Three solutions were used in this study:

30% solids	8 ml fill	Specific surface area=0.27 $m^2\ g^{-1}$
30% solids	4 ml fill	0.48 $m^2\ g^{-1}$
5% solids	4 ml fill	0.88 $m^2\ g^{-1}$

Although the sample with the lowest specific surface area has double the fill depth, the correlation between drying rate and specific surface area is good. It also demonstrates that cake depth does not significantly affect secondary drying kinetics.

In practice, the specific surface area is not a controllable variable. Although it is influenced by the degree of undercooling (related to cooling rate), this is likely to vary from vial to vial. Decreasing the cooling rate may be expected to lower the specific surface area and the secondary drying rate, but it would *increase* the optimum rate of primary drying by favouring the formation of large ice crystals with good connectivity. This discussion illustrates yet again that effective freeze-drying is best achieved by a compromise!

9.6 Effect of Temperature

It is clear from the state diagram that, if the temperature of a product is raised at a rate such that it increases faster than the increase in T_g resulting from water removal (secondary drying), then collapse, and

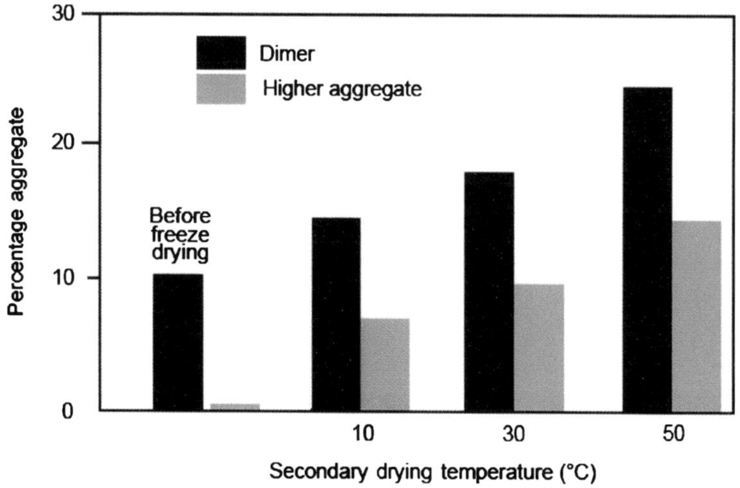

Figure 3 *The effect of secondary drying temperature on the aggregation of human serum albumin (HAS). The solution, containing 8% protein and 8% ethanol, was freeze dried at a primary drying temperature below −30°C; secondary drying was performed at the temperatures indicated. Reproduced from Pikal (unpublished data).*

potentially also aggregation and/or chemical degradation, will inevitably occur. This is graphically illustrated in Figure 3, which shows how product quality can be significantly and usually adversely affected by temperature abuse during secondary drying.

The histogram in Figure 3 shows that, even before the start of freeze-drying, 10% of the product was already present in its dimerised form, possibly due to some "rough" handling during the purification process. There was no indication of any more advanced aggregation at this stage. The product would have contained at least 20% of water, which would give it an estimated T_g' of *ca.* –25°C. Even with secondary drying performed at 10°C, considered by many to be a low temperature, HSA had become prone to aggregation, to a degree that would render it quite unsuitable for use in parenteral products. As expected, aggregation rapidly increased with rising temperature.

Surprisingly, even modern literature[13] carries statements such as: "During secondary drying, the product temperature has to be raised to the maximum tolerable temperature of the dried product. This raise (*sic*) can be done as quickly as can be technically achieved in the plant. Raising it more slowly does not make the product safer. When there is no more ice in the product, the final temperature can be applied." This advice is dangerously misleading as can readily be concluded from an inspection of Figure 3.

At a given temperature, the water content decreases with time but becomes rate limited, *i.e.* the water content/time profile will tend to a plateau. Raising the temperature will increase the rate of water removal until a new rate-limiting plateau is reached. For the record, Pikal *et al.* have noted that in the case of formulations containing mannitol in its normal crystalline form, no temperature effect is observed.[118]

The plateau effect demonstrates that products cannot be adequately dried to low water contents by carrying out secondary drying at low temperatures, unless very extended and uneconomical drying periods are employed. Increasing the product temperature is necessary but must be done cautiously. As water is removed, the glass transition temperature will rise and the product temperature may then be increased. More water will then diffuse out, again raising the glass temperature, and the product temperature may be further increased.

The plateau effect also demonstrates that, once the product has reached its final drying temperature, there is little value in extending the drying period beyond a few hours, and the practice of lowering the temperature for the final drying period is of no value if the product has already reached a steady state at a higher temperature.

The non-linearity of the Ficoll glass drying profiles in Figure 2 shows that (Fickian) diffusion in the sample, and not evaporation (from the surface) is the rate-limiting step. The secondary drying analyses also demonstrate that water is mobile in the glass, even well below the measured glass transition temperature. Its behaviour in the solid substrate is analogous to that of a plasticiser.

9.7 Softening and Collapse

Under reasonable operating conditions, residual unfrozen water can be removed from the glass only by raising the product temperature above the glass transition for a given period. It is thus almost impossible to avoid a minor degree of collapse, as shown in Figure 1. Good operating practice then demands that excessive heating be avoided during secondary drying. An idealised drying path is sketched in Figure 4.[120] Whereas primary drying is preferably carried out under isothermal conditions ($<T_g'$), the ideal secondary drying path tracks the midpoint between the almost-parallel T_g and T_s profiles. Drying can be terminated at some predetermined temperature (shown as *c* in Figure 4), depending on the desired final water content.

9.8 Operational Drying Protocols: Temperature Ramping and Stepwise Heating

In practice, it is hardly possible to control the drying process exactly as shown in Figure 4. A good compromise can however be achieved by judicious temperature/time ramping, always keeping the drying

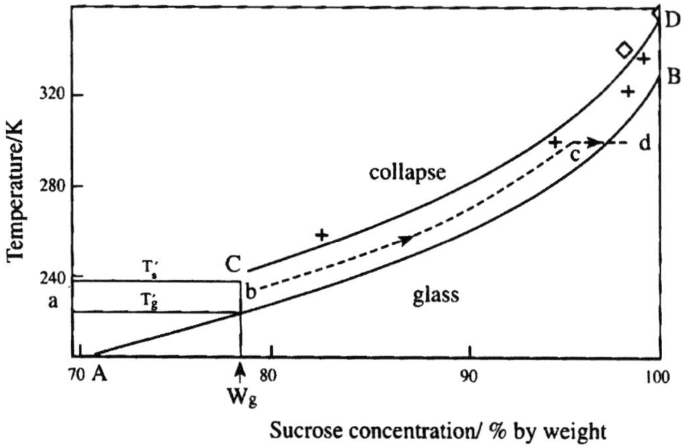

Figure 4 *Optimal freeze-drying pathways for the preparation of a "low-moisture" sucrose preparation. AB is the glass transition profile and CD the softening temperature profile as observed by DSC. W_g' is the composition of the maximally freeze-concentrated solution; ab is the ice sublimation at constant temperature; bcd is the ideal secondary drying path; (+) data from Shalaev;[121] and (◇) data from Shalaev and Franks[120]. For details, see text*

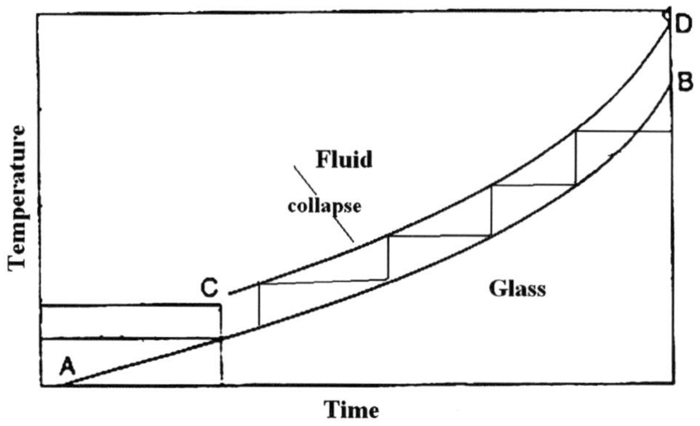

Figure 5 *Sketch of a practical secondary drying process, based on a stepwise temperature ramping, interspersed with thermal equilibration periods, between the limits of T_g and T_s. The process aims at a combination of economy of time with acceptable product quality and is based on the extension of the "ideal" drying protocol, shown in Figure 5*

temperature between the limits set by T_g and T_s. The temperature separation $\Delta T = (T_s - T_g)$ is typically of the order of 10 K (see Figure 4). The underlying principle of the method is illustrated in Figure 5a of Chapter 6, and a qualitative sketch of a recommended time/temperature protocol for secondary drying is shown in Figure 5, beginning at T_g' and ramping of the shelf temperature in steps of *ca.* + 5°C h^{-1}. Assuming that the formulation and fill depth do not exceed the recommended values, this protocol should normally allow enough time for sufficient water to be removed from the product to raise the glass temperature of the plug to the set shelf temperature. The plug will thus never be exposed to temperatures in excess of (T_g + 10 K) for unduly long periods.

CHAPTER 10

Some Practical Aspects

10.1 Scale-Up and ICH Manufacture

The removal of 99+% of water, at subfreezing temperatures, from the solution of a product that is potentially chemically and mechanically labile is a complex, stressful and costly operation. Once the physico-chemical parameters (*e.g.* T_g' and residual water content) have been established, the process optimisation is governed by the coupled heat and mass transfer rates, the operating scale and the particular equipment available. For small-scale formulation and process development, engineering considerations are rarely of consequence. As the operating scale and the associated level of automation (*e.g.* filling and stoppering) increase, these factors will become of ever-increasing importance.

The ultimate aim in the development of every freeze-dried product is full-scale manufacture, in line with market requirements. For many products, this can be in the range of 50,000–100,000 vials per batch. At this scale, many of the operations are automated, and consideration must be given to the practicalities of working at this scale. Also, as a product approaches the time when it will be administered to animals and, eventually, humans, health and safety factors enter the equation. The demands of the regulatory authorities will then have to be met. During the initial development of a freeze-dried product, it is quite typical to work on a laboratory scale with batch sizes of about 200 vials. Before committing to full-scale manufacture, process refinement and the production of sufficient samples for informal stability trials are carried out at an intermediate scale; it is simply a balance of sufficient production capacity, the cost in time and materials cost of "failed" batches. Pilot-scale refinement is conveniently carried in a drier of approximately 1 m^2 shelf area, equivalent to *ca.* 1200–1500 vials per batch. At this scale most ancillary operations, filling, capping (the application of aluminium overseals after stoppering), are carried out manually. When moving to a larger scale, many of these operations are carried out automatically, and this leads to further engineering considerations, *e.g.* the partial insertion of 50,000 stoppers.

In the pharmaceutical industry, the objective of the formulation scientist will typically be the manufacture of the three ICH stability test lots, as required for a regulatory filing. The rules of the ICH (International Conference on Harmonisation of Technical Requirements for Registration of Pharmaceuticals for Human Use) are shown in Table 1. Essentially they require the manufacture of three batches at a scale representative of the intended commercial scale, which are then subject to formal stability and quality trials, typically for a period of 12 months. With other dosage forms (*e.g.* oral tablets), this is usually taken as 10% of the proposed commercial production scale, but the interpretation for lyophilised product, *e.g.* intravenous parenterals, is not so clear, especially as the volume of water to be removed may impact upon the performance of the drying equipment. For example, would filling only one shelf of a 10-shelf drier be representative of the heat and mass transfer characteristics of a fully loaded drier? A more conservative approach is therefore taken, and in many cases commercial-scale lots are prepared for ICH stability trials. Apart from the inevitable delay that an ICH failure would introduce to the filing date, there is also a very real direct cost. Three ICH lots of a sterile product, carried out at a contract-manufacturing site, could easily run to $2 million. A full programme for scale-up and ICH manufacture of a sterile parenteral product could then easily amount to $4–6 million. Table 1 contains a typical list of operations that need to be carried out in order to complete such a programme.

10.2 Economics

Several economic analyses of freeze-drying are on record (*e.g.* Millman *et al.*[122]). They do, however, consider the process, as applied by the food industry. The conclusion is that the operation regime with the shortest drying cycle also gives the lowest production cost and the highest capacity. The recommendation has therefore been for large volumes. In view of the nature of the products and their robustness to the drying stresses, this seems a sensible strategy.

In pharmaceutical freeze-drying, on the other hand, importance is placed not on mass, but on activity/unit mass, and the parameters that govern survival of some biological activity in small volumes are totally different and more subtle than they are for a simple dewatering process.

Table 1 *A typical scale-up and ICH manufacture programme*

Safety issues
- Evaluation of operator safety
- Evaluation of waste disposal
- Evaluation of cleaning process

Packaging issues
- Selection and evaluation of vial, stopper and overseals
- Formal studies of extractable components of the packaging
- Development and qualification of stopper cleaning and sterilisation
- Development and qualification of vial cleaning, depyrogenation and sterilisation
- Line tests of vials and stoppers
- Optimisation of stopper siliconisation (where required)
- Purchase of necessary change parts to lines
- Fitting and qualification of change parts
- Container closure integrity (dye leakage)
- Vial headspace gas analysis (leak check)

Pilot-scale (1000 vials) development
- Cleaning verification
- Documentation for compounding and freeze-drying
- Tech transfer of compounding methods
- Transfer and qualification of compound-specific chemical analytical methods
- Agreement on bioburden, sterility and standard analytical methods
- Analytical methods transfer
- Process refinement: three batches at pilot scale

Transfer to full-scale sterile development batches
- Condenser water challenge
- Full-scale water batch (dummy runs to checkout *everything*)
- Cleaning verification and validation
- Media fills (confirmation of sterile manufacture)
- Finalisation of extractable volume and fill volume
- Preparation of documentation (filling and manufacturing works orders)
- Full-scale development batches (2, 3)
- Helium leak rate test

Manufacture of ICH stability lots
- Preparation of documentation (filling and manufacturing works orders)
- Manufacture of three ICH stability test lots

One also deals with a batch process with a reputation for high running costs; some consideration should therefore be given to the economics of the process.

Factors to be considered are as follows:

- Power (electricity)
- Rent/space
- Labour
- Ancillary processes
- Capital depreciation.

Freeze-drying can be shown to be an *energetically* inefficient process, but is it therefore necessarily expensive to run? A typical freeze-drier with a sufficient capacity to dry 100 kg of solution dispensed in 2-ml portions with a 5-mm fill depth and T_g' of −40°C has a power rating of 80 kW. For a "typical" freeze-drying cycle of 65 h, it would be possible to dry 40 batches annually. Energy (electricity) costs would be of the order of 0.5 cents per vial, whereas the purchase cost of the equipment, depreciated over 5 years, would amount to *ca.* 7.5 cents per vial. In practice, therefore, power costs tend to be insignificant compared to the value of "the product" and, when compared to the cost of (purified) raw materials, packaging and pre- and post-drying operations (*e.g.* filling and QA) make only a minor contribution to the overall M&A costs. The "fixed" costs include rent/buildings, manpower for servicing production, filling, packaging and depreciation. The "capital cost" and, hence, "depreciation" of the equipment can be a significant factor.

Other economic considerations that affect the manufacturing cost of the product include:

- Size/capacity
- Shift working patterns
- Length of cycle times/overall plant throughput
- The "fear factor" (is it prudent to risk one large freeze-drier run, where breakdowns can be catastrophic, or would two smaller freeze-driers be preferable?)
- Escalation of QC/QA costs with smaller batch sizes.

The final "cost" of a product will ultimately depend upon the accounting practices of the manufacturing company. Cycle time, however, is often singled out as an easily measurable parameter to which costs can be scaled. It is not unusual for management to exert pressure for a reduction in the cycle time. With any given formulation/fill volume/vial geometry combination, there exists an optimum freeze-drying cycle that can be estimated from measurements of physico-chemical parameters. Any attempts at corner cutting should be discouraged; they always run a risk of product quality downgrading. Saving one day in the production cycle will be a false economy if it results in halving of the product's shelf life.

Inspectors nowadays increasingly ask for cycle justifications, even for products that have been manufactured and sold without problems for a number of years. Rational formulation and process development and transparency in the documentation will result in a process that can be justified to both the company accountants and the regulatory authorities.

10.3 Vial Closure and Packaging

In the early stages of product development, vials are typically filled manually, and stoppers are partially inserted by hand, prior to freeze-drying. On a production scale, these processes are automated and line trials must be carried out to ensure that appropriate adjustment can be made to allow both vials and stoppers to move freely along the lines without incident and for stoppers to be accurately and reproducibly inserted prior to loading. The design of reliable and effective stoppering devices and mechanisms is of the utmost importance and this aspect has been treated in several monographs.[115]

If siliconised stoppers are used, typically a minimum silicone loading would be sought that would allow movement along the lines whilst not producing haze in the reconstituted final product. On a small scale, mismatching of stoppers and vials would typically be of little consequence. After stoppering, vials are quickly removed from the drier and aluminium overseal crimps are applied by hand. Stopper lifting would not generally be a problem, since sterility may not be an issue and any leaks will be quickly dealt with when the overseal is applied. In a large-scale process, even a small percentage of stopper lifting is a serious issue, as would also be any breach of sterility. Accordingly, the designs of both stoppers and vial necks may contain a number of features that are not immediately obvious but may lead to a potential mismatch.

One such design feature is the "blowback". This term refers to part of the vial neck design with two purposes:

(i) In the vent position, it serves to prevent "pop-out" and "drop-in".
(ii) In the closed position, it serves to prevent "pop-out" before sealing.

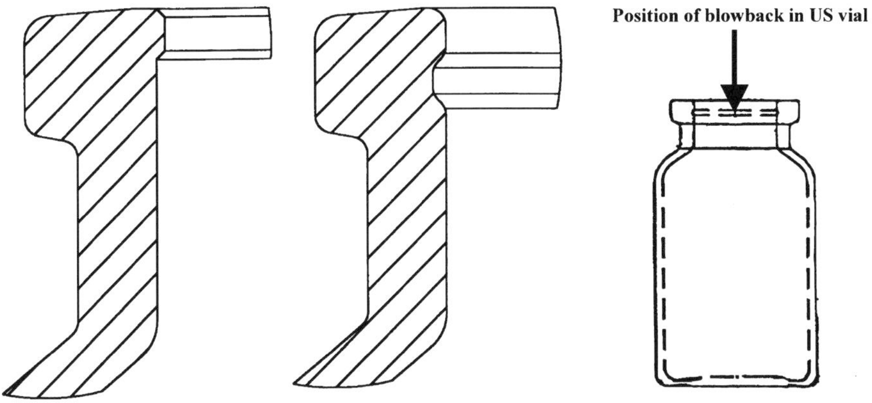

Figure 1 *Different styles of blowback features in vial necks. Left: European style; centre: US style and right: position of US style blowback*

There are two styles of blowback in vial necks, as illustrated in Figure 1. The design on the left is the European ridge style. The centre diagram shows the US groove style design, and the right-hand diagram shows the position of the US style blowback in the US vial. Although the sealing surface of the stopper is the underside of the flange, care should still be taken to match stopper and vial designs. Figure 2 shows typical European and US style stoppers, designed for use with appropriate blowback features.

Stopper design also contributes to the overall resistance to ice sublimation. Figure 3 shows how the sublimation of ice from a frozen lactose solution, using identical vials, is affected by different stopper designs. Thus, vials, fitted with stoppers of type B, gave a 15% reduction in the rate of ice sublimation. The final data point on the "Stopper B" data set is in fact a third design that affects the rate even more markedly.

The effect of stopper design on the sublimation rate has been studied in detail,[13] and examples are shown in Figure 4. By affecting the sublimation rate, stopper design will also affect the product temperature. Changing stoppers may be particularly problematic if the traditional, but not necessarily recommended, practice of using the ice sublimation rate to maintain a low product temperature is used.

10.4 Process Monitoring and Control

A reliable control of the product temperature throughout the process is of the utmost importance, especially during the primary drying stage. In practice, such control can only be exercised through changes in the shelf

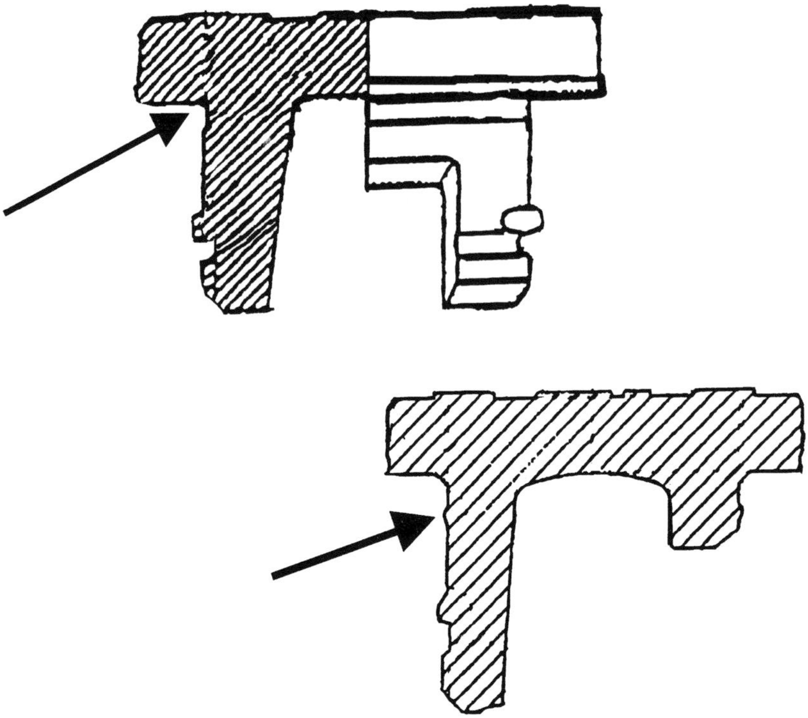

Figure 2 *Stopper designs to match European (top) and US (bottom) style blowback*

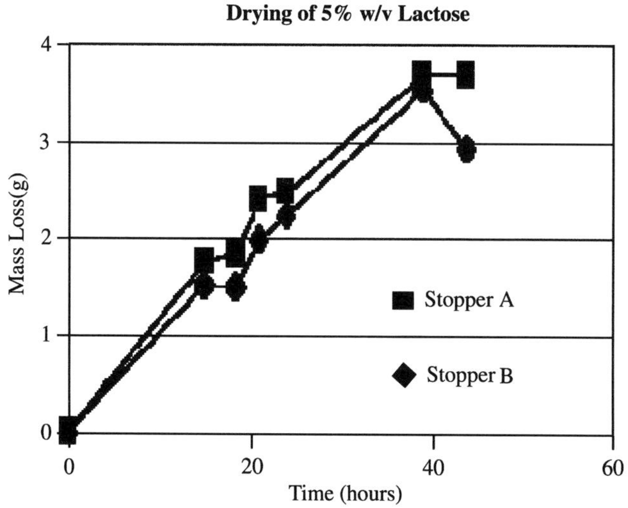

Figure 3 *The effect of stopper design on the sublimation rate from a frozen 5% lactose solution*

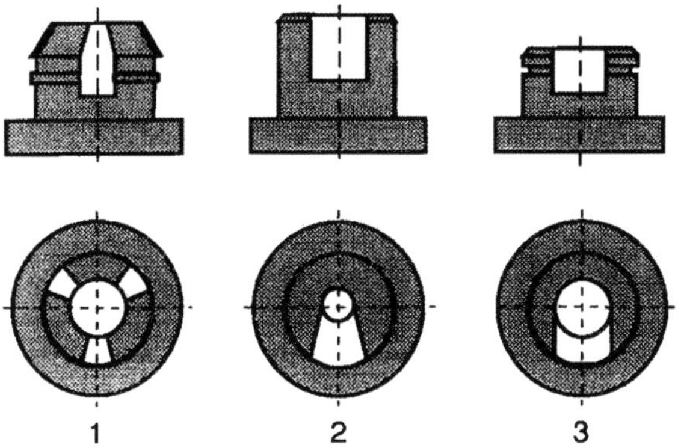

Figure 4 *Amount of ice sublimed at the same temperature in 3 h, scaled to unstoppered vials (= 100%); stopper 1 = 77%, stopper 2 = 75% and stopper 3 = 66%. Data from Willemer*[13]

temperature and/or the chamber pressure. Preferably, the product temperature should be monitored directly by probes placed in vials at predetermined locations within the freeze-drier. Probes consisting of thin wire thermocouples are preferable to resistance thermometers (RTD sensors), but any probe must remain in close contact with the product throughout the drying cycle. However, any probe will liberate heat; hence, product temperature readings will not correspond exactly to the temperature of the vast majority of vials that do not carry a probe. In practice, the heating effect of the thin wire thermocouple is negligible, compared to that of the RTD probe; also sample contact is more easily maintained. It should also be remembered that the presence of any solid temperature sensor in a vial will increase the ice nucleation probability *in that vial*. The detection of a freezing event is therefore not representative of the average freezing temperature, taken over all the vials in the freeze-drier.

The temperature of the advancing ice front can be determined using a barometric temperature measurement, via a pressure rise test (PRT) (see later), using a water vapour–temperature diagram, as shown in Figure 5. In practice, this requires an interruption of the cycle and lacks the convenience of the continuous output from a thermocouple.

In some commercial equipment, a conductimetric method is employed to measure the so-called eutectic temperature. In practice, where supersaturated solutions are the rule, eutectic temperatures are rare, and where they do occur, they are unrelated to the physical state of the drug substance. Such measurements are of little value other than indicating a change in conductance with temperature. In this role, they might give a warning of abnormal conditions.

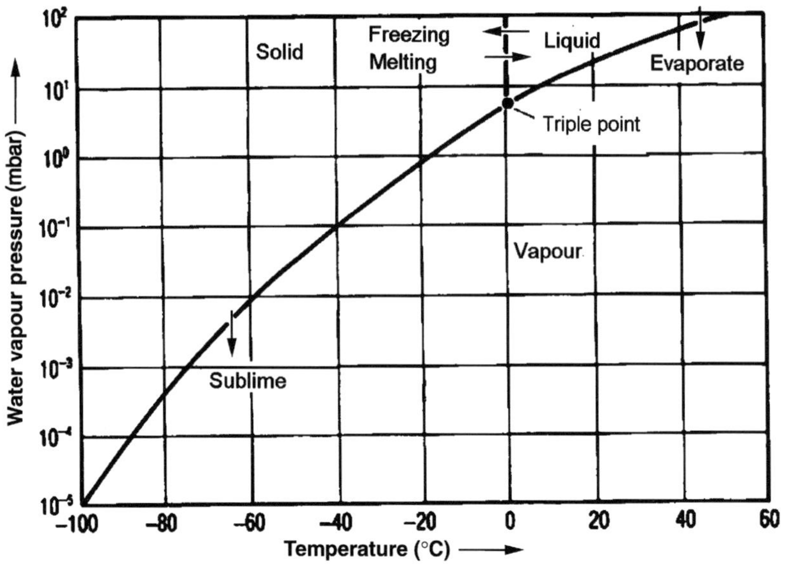

Figure 5 *Water vapour pressure–temperature relationship*

Apart from the product and shelf temperatures, the condenser temperature should also be monitored. This would normally be set to its lowest achievable value. A rise in the condenser temperature at some stage during the drying process is a danger sign that the condenser is unable to cope with the mass transfer rate of water vapour.

The other process variable that needs to be controlled and monitored is the chamber pressure. Older instruments use a heat conductivity vacuum gauge to determine pressure. Such gauges measure the electrical energy needed to maintain a constant temperature. They are typically calibrated against air, so that the readings need to be corrected for pure water vapour. Tables are available for this purpose, although an approximate value can be obtained with a conversion factor of 0.65. Membrane differential gauges, or capacity gauges, offer high resolution and reproducibility and measurements that are independent of the gas type. The principle makes use of a membrane, sealed against a fixed low pressure, that is part of a capacitor and the capacitance changes caused by a deflection of the membrane are determined.

The PRT is a popular method of determining the completion of ice sublimation. The main chamber is isolated from pump and condenser, and the rate of pressure increase is measured. In its simplest form, which suffices for most purposes, sublimation is deemed to be complete when the PRT value is equivalent to the background leak rate of the equipment, typically 1.5–3.0 Pa/min.

10.5 Stability Issues

Much has been, and is still being, written about product stability. Hardly a month goes by without a symposium devoted to regulatory/legal aspects and/or measurement procedures to satisfy the various demands placed on the manufacturer of biopharmaceuticals. This book is not the place for a detailed treatment of legal stability issues; they are adequately covered elsewhere.[123,124] Let us therefore confine this discussion to important generalities and guidelines. The approaches that must be made to satisfy the American regulatory demands differ from those operating within the European Union (EU). The US Food and Drug Administration (FDA) requires that the freeze-drying process should be properly defined, validated and controlled. The EU, on the other hand, operates by a series of "Directives", to be framed by each member country into national legislation, with an aim of harmonisation.

Before attending to the finished product, many preparatory regulatory demands must be addressed to establish adequate stability. They cover the whole process, specifically the following aspects:

- product formulation, including stability of all ingredients;
- holding time and holding environment prior to freezing;
- filling procedures and sterility, vial stoppering;
- drying cycle control and validation.

Consider, as a typical example, the closure systems that are proposed for a given product. To pass the regulatory demands, these systems will need to be tested for extractable and leachable substances that would end up as contaminants in the final product. Extractables arise from individual components of the packaging systems under appropriate solvent and temperature conditions. Leachables are compounds that migrate from the container/closure system of the formulated drug product under normal conditions of use or during stability studies.

The finished product will be subjected to inspection and rigorous testing for identity, uniformity, residual water content, stability, sterility and potency. In addition, all analytical techniques employed in testing these attributes will themselves have been subjected for reliability, reproducibility, experimental uncertainty limits. The biotechnological revolution has resulted in the appearance of ever more refined and sensitive analytical techniques, mainly novel types of spectroscopy and coupled techniques, based on mass spectrometry, known usually by complex acronyms, *e.g.* MALDI-TOF-MS (*M*atrix-*A*ssisted-*L*aser-*D*esorption-*T*ime-of-*F*light-*M*ass *S*pectrometry). Some of the available analytical procedures are treated in more detail in the next chapter.

It is beyond the capability of all but the largest pharma companies to bring to bear the complete battery of available analytical techniques in efforts to satisfy the regulators. A careful analysis must precede the selection of a combination of those techniques that will provide acceptable answers to specific questions, such as toxicity, aggregation, the nature of trace contaminants or loss of potency.

CHAPTER 11

The Dried Product

11.1 Physical State: Crystalline or Amorphous?

The primary aim of a commercial pharmaceutical freeze-drying operation is to deliver a marketable product that possesses an acceptable shelf life under certain specified conditions, but one that can also readily be reconstituted at the point of use. This aim is achieved by attention to a combination of formulation and processing details. As the foregoing discussions would have made clear, a choice of the physical state of a freeze-dried preparation open to the product developer is in most cases quite limited because of practical problems associated with crystal nucleation and growth, or the lack thereof in real time. The options are further restricted in the preparation of biopharmaceuticals, especially protein-based products, because the vast majority of their formulated preparations cannot be completely crystallised. The combination of crystalline and amorphous states in a dried preparation usually gives rise to extremely complex phase behaviour. Figure 1a in Chapter 4 represents a typical equilibrium phase diagram of the simplest type of solid/liquid binary aqueous mixture (KCl–H_2O), where both components crystallise with an ice-solute eutectic point at −11°C and 20% w/w KCl. The situation becomes slightly more complicated when the solute crystallises as a hydrate. This is shown in Figure 1 for the system NaCl–H_2O, where the crystallising solute species is a metastable dihydrate $NaCl.2H_2O$, which decomposes on heating at a peritectic temperature of 0°C into NaCl and water.

By contrast, Figure 2a shows a hypothetical isothermal cut through a state diagram for a ternary aqueous mixture in which water and one solute (*e.g.* glycine) can crystallise under certain conditions, but the other solute (*e.g.* sucrose) will invariably vitrify.[125] Results from a detailed investigation of the water–glycine–sucrose system, including

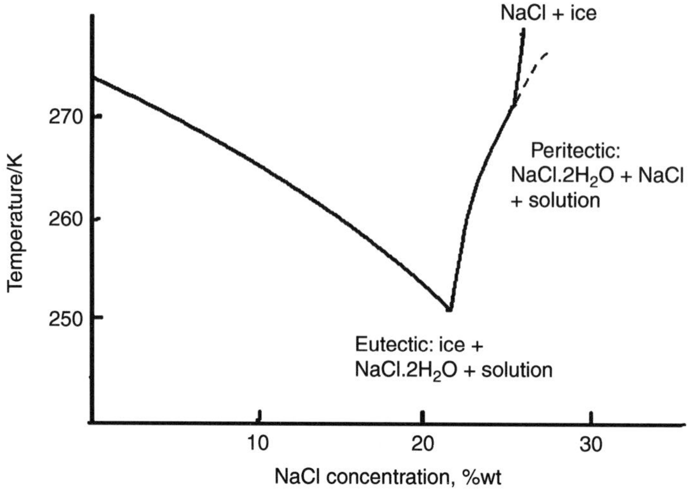

Figure 1 *Solid/liquid phase diagram of the NaCl–H_2O system. The eutectic point occurs at 252.15 K (76.89% NaCl) and the peritectic point at 273.28 K (26.3% NaCl)*

all theoretically possible crystalline and amorphous phases/states, highlight the complexities that may be encountered during the freeze-drying of such multicomponent mixtures; they are indicated in Figure 2b[126] in the form of a full temperature–composition state diagram. The complete cooling path of mixtures with compositions P and N can be traced, through the notional eutectic points onto the ternary glass, and the final end of freeze-concentration. This representation does, however, look

---

Figure 2 *(a) Projection of a temperature–composition state diagram for a hypothetical ternary water/solute 1/solute 2 system on a composition triangle. The e_1, e_2 and e_3 points correspond to the three binary eutectics ice-1, ice-2 and 1-2, and point E corresponds to the ternary eutectic ice-1-2. Curves aa′ and bb′ are isotherms for primary ice crystallisation at temperatures T_1 and T_2. Curve kk′ traces the composition of the freeze-concentrated solution after the completion of primary ice crystallisation. Compositions within the field e_1Ee_3A describe primary solute 1 crystallisation, and points within e_1Ee_2B to primary ice crystallisation and e_3Ee_2C to primary solute 2 crystallisation. For a full description and significance of Figure 2a, the reader is referred to Shalaev and Franks.[125] (b) Complete temperature/composition phase and state diagram for the ternary system water–sucrose–glycine. Experimental data are shown by triangles. The space bounded by $T_{ma} - E_{gs} - T_{ms} - E_{wg} - E_t - E_{sw}$ is the theoretical ternary equilibrium phase diagram. The dark space, bounded by the three glass transitions of the pure components: θ_s, θ_w and θ_g is the region of supersaturation. Subscripts w, s and g refer to water, sucrose and glycine, respectively; θ'_{ij} are the three binary T'_g values, E_{ij} are the three binary eutectic points and E_t is the ternary eutectic point. PQG and NOS are cooling paths for two representative mixtures (P and Q), accompanying freeze concentration. Redrawn from Suzuki and Franks, 1993.*

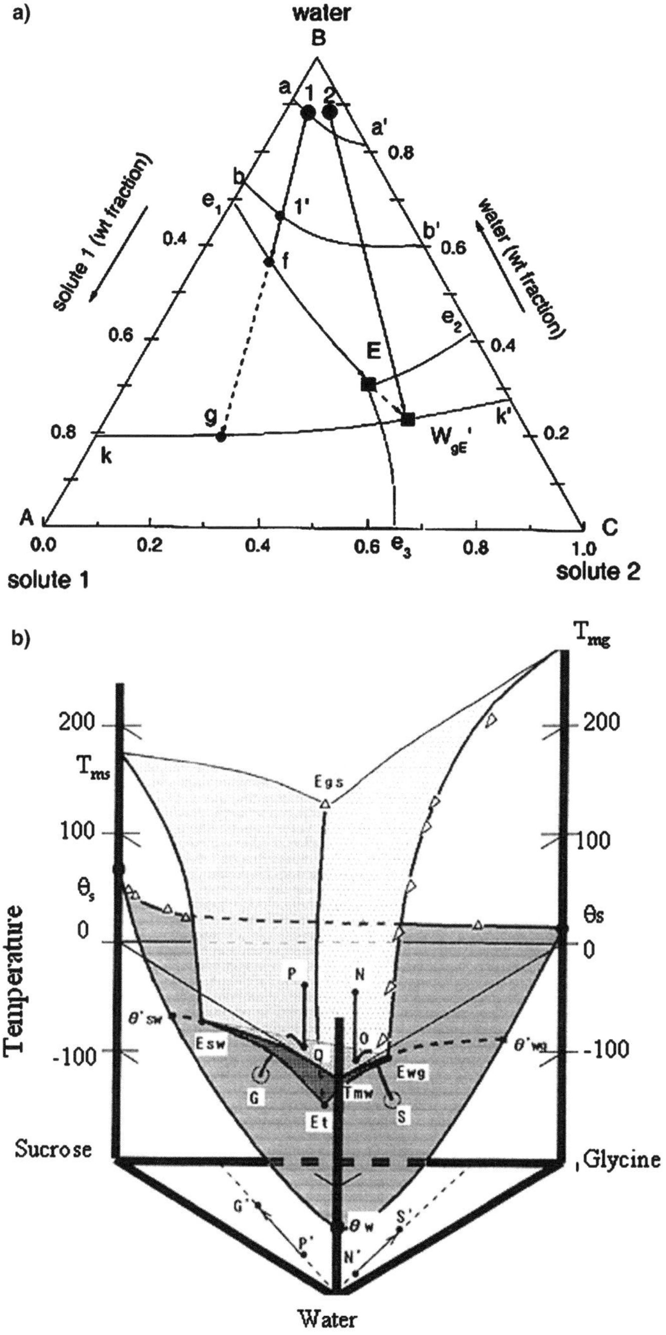

a)
water
B
a
1
2
a'
0.8
b
e1
1'
b'
0.4
0.6
f
solute 1 (wt fraction)
water (wt fraction)
e2
0.6
0.4
E
g
k'
0.8
0.2
k
WgE'
A
C
0.0
0.2
0.4
0.6
e3
0.8
1.0
solute 1
solute 2
b)
Tmg
200
200
Tms
Egs
100
100
θs
θs
0
0
Temperature
P
N
θ'sw
O
O
θ'wg
-100
-100
Esw
Ewg
G
Tmw
S
Et
Sucrose
Glycine
G'
θw
S'
P'
N'
Water

somewhat complex, because it has to combine the solid–liquid phase relationships of three binary mixtures and the ternary systems, in addition to the "state" diagram, which represents the T_g data for the four systems. In practice, therefore, the isothermal sections, such as the one in Figure 2a, tend to be used to obtain solubility, glass and other physical data.

Because of the reluctance of many substances to crystallise in real time from frozen aqueous solutions, biopharmaceutical therapeutic agents will generally be marketed as glasses. When this fact first dawned on technologists in the food and pharmaceutical industries,[71,127] it was believed (assumed?) that the glass transition temperature T_g formed a clear dividing line between stability and instability. As time went by, however, it came to be realised that the behaviour of multicomponent, water-soluble glasses is complex, and that "absolute" or commercially acceptable stability is influenced by factors beyond the measured T_g. A proper appreciation by product and process developers of the complex materials science of water-soluble glasses is nowadays considered as essential, although some unresolved mysteries still remain,[128] see also Section 11.10.

Relative advantages of crystalline *versus* amorphous (glassy) preparations have already been discussed in Chapter 7. Because the (ideal) crystalline state *always* has a lower energy than any amorphous state, it has often been suggested that this is the form to be aimed at for maximum long-term stability. Indeed, many solid dosage forms can be successfully dried into stable crystalline preparations. It must be noted, however, that rigours of a drying process may produce subtle changes in molecular structure, such as, for instance, the racemisation of formerly chiral compounds, oxidation (vitamins), decarboxylation (moxalactam), oxidation (aspirin) or photochemical decomposition (gibberellins). Drying may also lead to the formation of unexpected hydrates and their equally unexpected decomposition, to polymorphic changes, or to mixtures of polymorphs,[129,130] thus changing the physical properties of the dried product. Mannitol and lactose are particularly prone to such changes during drying and also during subsequent mechanical processing.

11.2 Essential Materials Science of Crystalline and Amorphous Products

The production of crystalline materials is governed by the laws of crystal nucleation and growth, which are reasonably well established (see Chapter 4). Complexities arise in multicomponent mixtures, where such rates for different components may differ by orders of magnitude. Problems may also arise from molecular interactions that inhibit (poison) the crystallisation of individual components. This phenomenon has, however, been usefully employed in a full characterisation of the glass properties of mannitol, when its crystallisation is interfered with (poisoned) by the addition of low concentrations of sorbitol.[57]

In the production of solid pharmaceutical preparations, it is important to be aware of the identities of polymorphs that might be formed during the drying process, or during any later processing steps, *e.g.* milling, compression, extrusion.[131] The pressure-induced formation of specific solvates with superior stabilities has also been reported, *e.g.* for paracetamol.[132]

In situ hydrate crystallisation, facilitated by residual water under certain storage conditions can, however, also be of practical value. This has been described by Aldous *et al.*[100] for freeze-dried α,α-trehalose and raffinose. The trehalose preparation contained 8% residual water, which gave it a T_g of 31°C. After overnight storage at 82°C, a T_g of 79°C was recorded, followed by a crystal-melt endotherm at 97°C, characteristic of trehalose dihydrate. The results are shown in the form of differential scanning calorimetry (DSC) heating traces in Figure 3. Similar experiments, performed on raffinose, containing 3.7% residual water, produced qualitatively identical results. Overnight storage a few degrees below the melting point of the pentahydrate, led to the disappearance of a previously

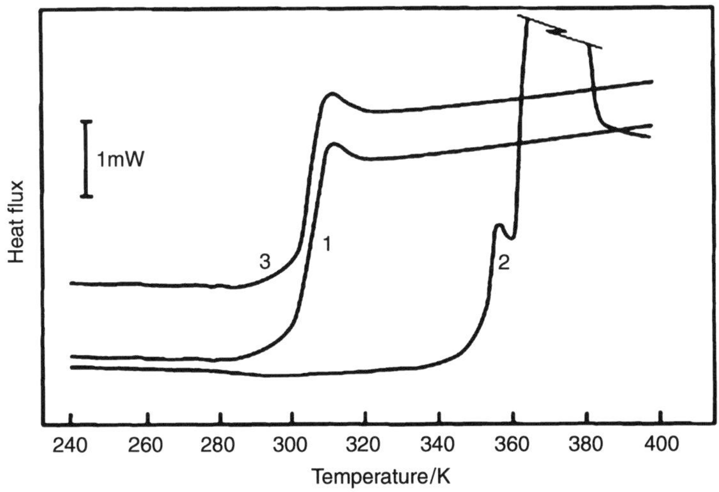

Figure 3 *Successive DSC heating scans for air-dried trehalose containing 8% residual water. 1: first heating scan; 2: heating scan after overnight exposure to 355 K (82°C) and recooling to 2303, and 3: heating scan after melting and recooling to 230 K. Note the shift in T_g accompanying crystallisation. Reproduced with permission from Aldous et al.*[100]

recorded T_g at 63°C, which was replaced by the melting endotherm of raffinose.$5H_2O$. Thus, overnight storage had resulted in the complete sequestration of water from the amorphous phase by the crystallisation of the pentahydrate. This type of water removal has been termed "self-stabilisation", because it can be employed to protect the stability of labile pharmaceuticals against inadvertent exposure to water vapour during processing and/or long-term storage.

Compared to crystalline materials, the production and handling of amorphous substances are subject to serious complexities. Whereas the formation of crystalline materials can be described in terms of the phase rule, and solid–solid transformations (polymorphism) are well characterised in terms of pressure and temperature, this is not the case for glassy preparations that, in terms of phase behaviour, are classified as "unstable". Their apparent stability derives from their very slow relaxations towards equilibrium states. Furthermore, where crystal structures are described by atomic or ionic coordinates in space, that which is not possible for amorphous materials, by definition, lack long-range order. Structurally, therefore, positions and orientations of molecules in a glass can only be described in terms of atomic or molccular distribution functions, which change over time; the rates of such changes are defined by time correlation functions (relaxation times).

It is not immediately obvious where the glass transition fits into this picture of the amorphous state. Few reports exist of structural comparisons between crystalline, glassy and fused states of the same substance. One such study for glucose, based on neutron scattering, led to the conclusion that structurally, the glass can be best modelled as a deformed crystal, in which the unique atomic separations and hydrogen-bond angles have become somewhat distorted during fusion and quenching into the glassy state, *i.e.* partially arrested crystallisation.[79] Despite the structural similarities of the crystal and the glass, their thermomechanical properties are dramatically different.

It was earlier mentioned that in an amorphous material, cooled to within its narrow vitrification temperature range, molecular relaxation rates decrease by several orders of magnitude, so that physical and chemical changes become very slow, hardly accessible within the time scales that govern ordinary experimentation. It is nevertheless important to be aware that such changes *do* occur, even though relaxation times may increase to weeks, months, years or centuries. In practice, below T_g, molecular relaxation times become too long for equilibrium to be established on an experimental time scale. This is related to a drastic reduction in the number of accessible configurations, which a system can adopt as the temperature is lowered. If $W(T)$ is the probability of a transition, leading to a lower energy state, and z is the number of possible configurations (or the size of a domain in which the transition takes place), then

$$W(T) = A \exp(-z\,\Delta U/kT) \quad (1)$$

where ΔU is the energy barrier opposing the rearrangement. The critical size of the domain (or the lowest number of possible configurations) z^* is related to the *configurational entropy* S_c by

$$z^* = \mathrm{N}s_c^*/S_c \quad (2)$$

Here N is the Avogadro number and s_c^* is the *critical* configurational entropy, the minimum value of which is given by k ln 2, corresponding to only two possible orientations, such as those existing in imperfect crystals of carbon monoxide (C–O and O–C) and nitric oxide.

The average probability of a transition is given by A exp $(-C/TS_c)$, and if the constant C is not equal to zero, then the situation where $S_c = 0$ cannot be attained by a cooling process of finite duration. In the limit of low temperature, as $S_c \rightarrow 0$, so $T \rightarrow T_o$, which is a notional temperature corresponding to infinite stability; it is also referred to as T_K, the "Kauzmann temperature", in recognition of the seminal publication, which laid the groundwork for subsequent studies of the glassy state.[133] Therefore, $T_g \neq T_o$, but $T_g > T_o$, and this difference increases with the magnitude of C, *i.e.* with ΔU. Thus, *real* glasses have residual configurational entropies that increase with the difficulty of rearrangements or of translational motions within the molecular ensemble. S_c can be obtained by calorimetry, see later. In the present-day literature, T_o is commonly referred to as the "temperature of zero mobility". For many "real materials", such as ceramics, polymers, alloys, *etc.*, it has been found empirically that $T_g/T_o = 1.29 \pm 11\%$. For formulated pharmaceutical glasses, which invariably contain components of different sizes and shapes, the situation is of necessity more complex, because one then deals with a range of molecular mobilities, some of which, especially the residual water, may well affect the overall stability of the product.

Under conditions where $[(T - T_o)/T_o] << 1$, it can be shown that

$$W(T) = A \exp[-B/(T - T_o)] \quad (3)$$

where A and B are empirical constants; their significance is explained later. This is one form of the well-known Vogel–Tammann–Fulcher (VTF) equation,[134–136] which adequately describes diffusion and viscous flow, *although not necessarily in the neighbourhood of* T_g.

Although the amorphous, glassy state can confer a degree of kinetic stability upon an otherwise unstable system, a glassy state alone does not universally guarantee an acceptable long-term stability. The higher molecular mobility in the amorphous state, relative to the crystal, can lead to processes such as crystallisation or chemical degradation. Angell[137] first suggested that glasses can generally be classified into two types, according to the temperature dependence of their thermomechanical behaviour. He introduced a so-called fragility parameter, which is now often employed to describe the temperature dependence of molecular motions in an amorphous material, and the estimation of the fragility parameter is potentially useful for predictions of long-term product stability (shelf life). Thermal analysis, used to estimate the temperature dependence of the mean relaxation time τ, or the viscosity η, offers a practical method for estimating glass fragility (see later).

Figure 4 shows the extremes of the two types of behaviour: a scaling temperature was chosen (the notional glass temperature) at which the viscosity η reaches 10^{12} Pa s. The strong liquids display Arrhenius behaviour, *i.e.* ln η is linear in T/T_g. By contrast, the viscosities of fragile fluids decrease more rapidly at temperatures above T_g; this behaviour is described by the VTF equation. SiO_2 and Ge_2O are typical examples of strong fluids; they have tetrahedrally coordinated structures (like ice!), remnants of which apparently persist when the glasses are heated across the glass transition. Curiously, it has been found that, according to the fragility classification, globular proteins are also strong fluids. In fragile fluids, on the other hand, any residual structural order in the glass is rapidly destroyed above T_g. This is the case for substances

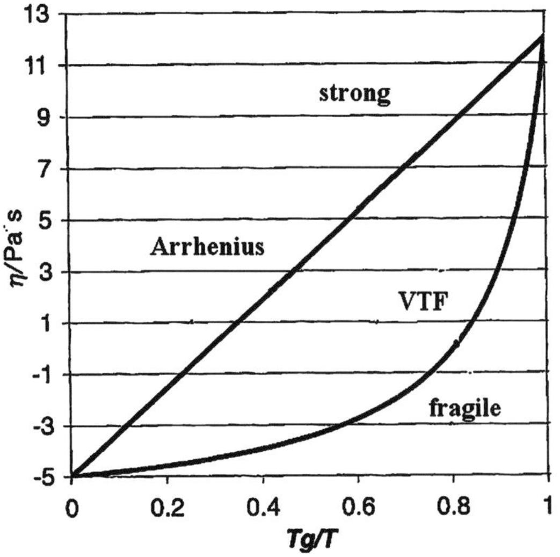

Figure 4 *Temperature dependence of viscosity: limits for strong and fragile fluids*

in which such residual order is due to hydrogen-bonded networks or van der Waals interactions. Figure 5 shows a selection of ln η versus T/T_g curves for pure polyhydroxy compounds (PHCs) and freeze-concentrated aqueous solutions. Also shown in Figure 5 is a band of viscosity values (in the neighbourhood of 10^8–10^7 Pa s), where viscous flow can first be directly observed in real time.[121]

The VTF equation for the temperature dependence of viscosity can also be written as

$$\eta = \eta_o \exp(DT_o/[T - T_o]) \tag{4}$$

It describes the temperature dependence of viscosity in an undercooled liquid, where η_o and T_o are constants, and D is the fragility parameter. Undercooled liquids with large D values (>50) are referred to as "strong", and those with low D values (<10) as "fragile". Molecular mobility behaves in a similar way and τ, the mean relaxation time, may also be described in a similar fashion:

$$\tau = \tau_o \exp(DT_o/[T - T_o]) \tag{5}$$

D is related to the energy of activation ΔH^*, and T_o may be equated with T_K, the Kauzmann temperature, or the isoentropic point between an undercooled liquid and a crystal. The constant τ_o corresponds to the high temperature limit of unrestricted mobility, which is often taken to lie 17 orders of magnitude below the experimental τ value. When $T_o = 0$, the VTF equation reduces to the Arrhenius equation, with D

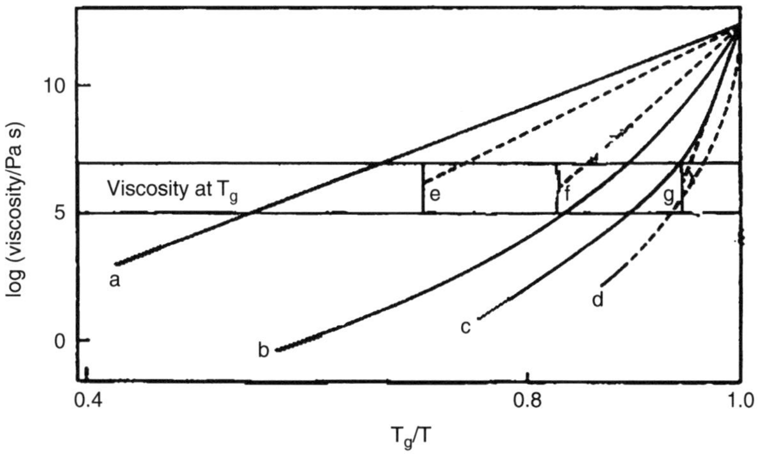

Figure 5 *Viscosity/temperature plots of amorphous materials to illustrate the impact of fragility: (a) silica, (b) glycerol, (c) 80% sucrose solution, (d) anhydrous sucrose, and maximally freeze-concentrated solutions of (e) ovalbumin, (f) glycerol and (g) sucrose. For explanation, see text*

proportional to ΔH^*, and when $T_o > 0$, ΔH^* becomes temperature dependent. D and τ_o are seen to be the two constants B and A in Equation (3). Non-Arrhenius, fragile behaviour in an undercooled region is explained by co-operative molecular motions, and the Adam–Gibbs equation:

$$\tau = \tau_o \exp(C'/TS_c) \tag{6}$$

where C' is a constant, uses the concept of configurational entropy (S_c) to describe the co-operatively rearranging domains. The Adam–Gibbs equation is important because it links undercooled liquid dynamics with the thermodynamic function S_c.[138] The temperature dependence of S_c is given by

$$S_c = \int_{T_2}^{T} \left(\frac{C_p^{conf}}{T} \right) dT \tag{7}$$

C_p^{conf} is the configurational heat capacity, *i.e.* the difference in C_p between the liquid (glass) and crystal states, and T_2 is the temperature at which the configurational entropy reaches zero, T_K. A modified Adam–Gibbs model describes non-equilibrium relaxation in the glassy state. A fictive temperature $T_{f(T,t)}$, is introduced, which represents the notional temperature at which the equilibrium glass enthalpy equals that of the non-equilibrium glass; Equation (7) then accounts for the temperature and time dependence of the non-equilibrium state. The Adam–Gibbs–Vogel equation

$$\tau_{(T,t)} = \tau_o \exp\left(\frac{DT_o}{T(1 - T_o) / T_{f(T,t)}} \right) \tag{8}$$

is a modified version of Equation (6) and reduces to the VTF equation for a quasiundercooled liquid, $T_{f(T,t)}$ is then equal to T.

The VTF and the Adam–Gibbs equations are useful because they permit the use of thermodynamic data, obtained by calorimetric methods, to describe molecular motions in undercooled liquids and glasses. Most investigations of pharmaceutical solids rely on calorimetric measurements, and for that reason their discussion is deferred until the fundamentals of the methods have been described in the following section.

11.3 Essentials of Thermoanalytical Techniques

For the development of effective freeze-drying cycles, the knowledge of glass transition temperatures of freeze-concentrated solutions, T'_g, is a *sine qua non*. On the other hand, the glass transition temperature of dried products, T_g, is indicative of their long-term stabilities. Glass transitions are conveniently measured by thermoanalytical methods, *e.g.* DSC, thermomechanical analysis (TMA) or thermogravimetric analysis (TGA). Long-term stability can, however, be more effectively measured by isothermal calorimetry. Because of their extensive use in pharmaceutical technology, it is appropriate at this stage to introduce a brief discussion of thermal scanning techniques and their place in the estimation of stabilities of freeze-dried pharmaceuticals.

DSC measures the difference in the heat flow to a test sample compared to that of an inert reference material (over the temperature range of relevance to freeze-drying, this is usually an empty matched sample holder or dry sand). A constant heating or cooling rate is applied and power is fed to the sample (or reference) so as to maintain a zero temperature differential. The calorimeter produces an output of heat flow against temperature. Since a constant heating/cooling rate is employed, the heat flow axis is in fact the specific heat C_p. Several types of transition can be observed: thus, a glass transition is detected as a discontinuous increase (during heating) in the specific heat over a relatively narrow temperature range. A first-order phase change is detected as a peak or trough, depending on the sign of the heat absorbed/released and on the configuration of the particular instrument used. A typical DSC heating scan is shown in Figure 6. The glass transition is identified by the step increase in the heat flow (corresponding to the specific heat). First-order thermodynamic phase transitions, such as crystallisation, many solid-state polymorphic transitions and melting, show up as exothermic or endothermic peaks, respectively.

As regards the study of freeze-drying in all its aspects, temperature scanning methods have become the most widely used tools.[139] Usually a controlled rate temperature programme – cooling/heating/cycling – is applied to the test sample and compared with a reference standard. A wide variety of physical/chemical process information can be obtained from changes in energy:

- rheology: elastic modulus, glass transition, plastic deformation, viscous flow;
- thermal conductivity;
- phase changes and phase diagrams: solid/solid, solid/liquid, solid/gas, *etc.*;
- rate constants of physical/chemical/biological processes.

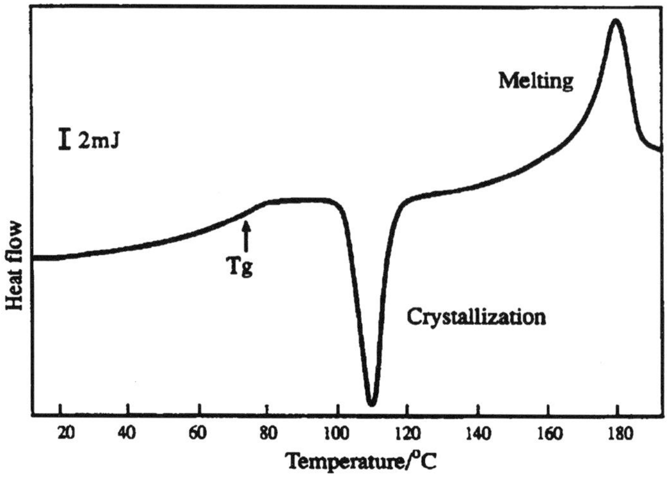

Figure 6 *Heat flow DSC scan for freeze-dried sucrose, containing 0.5% w/w residual water. Above* T_g*, the sugar becomes vulnerable to more or less rapid collapse and/or sucrose crystallisation, followed by melting at its equilibrium* T_m

The appearance of the DSC signal (resolution, shape and intensity) depends on:

- scan rate;
- sample mass;
- (solution) concentration and fill volume;
- type of sample cell: open, sealed, pin hole;
- instrumental sensitivity;
- purge gas: helium, argon, nitrogen, dry air;
- reference material: empty cell, sapphire, dry sand, *etc.*

Figure 7 represents cooling scans, showing both freezing and eutectic crystallisation exotherms of aqueous β,β-trehalose solutions at identical scanning rates (5 K min^{-1}).[163] They illustrate the fundamental principle of spectroscopy that resolution (signal width) is improved at the expense of signal intensity. Similarly, an increase in the scanning rate increases the recorded intensity but decreases the ability to discriminate between closely spaced transitions.

DSC, as applied to the analysis of freeze-drying processes, can be made more informative by the employment of variants, such as thermal cycling and/or annealing. An illustration is provided in Figure 8. The aim of the investigation was to resolve the origins of *two* apparent glass

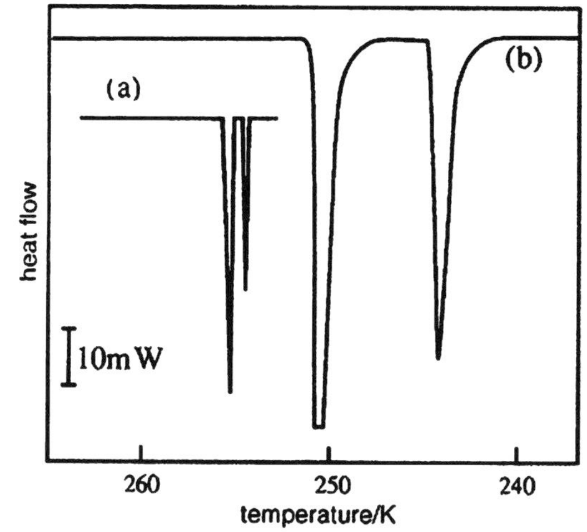

Figure 7 *DSC cooling traces (5 K min^{-1}) for aqueous β-trehalose solutions: (a) 8% w/w and (b) 21% w/w, showing freezing and eutectic crystallisation exotherms. Note the effect of concentration on the nucleation temperatures. Reproduced with permission from Roberts and Franks*[163]

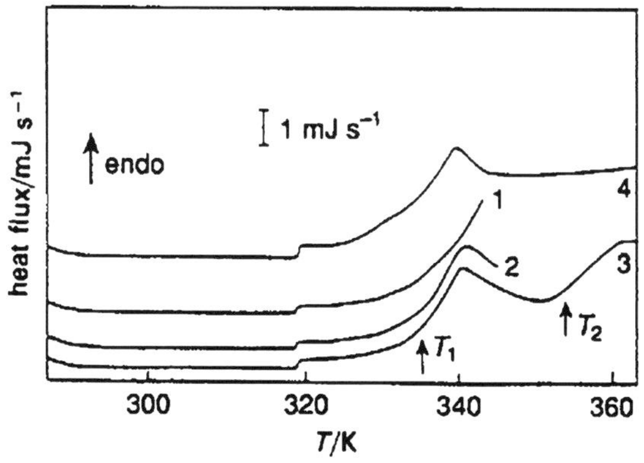

Figure 8 *DSC heating scans for freeze-dried sucrose, showing the effects of thermal cycling; heating rate: 10 K min^{-1}, sample mass: 8.232 mg. Scan 1: first heating, terminated at 364 K; Scan 2: second heating, terminated at 345 K (thermal history: heating from 280 to 346 K – cooling to 280 K – heating to 345 K); Scan 3: third heating, terminated at 365 K (thermal history 280–346–280–345–280–365 K); and Scan 4: fourth heating to 370 K (thermal history 280–346–280–345–246–280–345–280–370 K). For clarification, Scans 1, 2 and 4 are offset relative to Scan 3. Reproduced with permission from Shalaev and Franks*[120]

transition endotherms, frequently observed in DSC heating scans.† By subjecting a sample to several scans, interrupted at specific temperatures and subjected to specific annealing periods, it became apparent that one of the observed transitions was completely reversible for both frozen and freeze-dried materials, whereas the transition at a higher temperature was irreversible, *i.e.* it only occurred once, during the initial scan. In addition, this transition only occurred in disperse systems, such as frozen solutions or freeze-dried cakes, but it could never be observed for homogeneous samples, *e.g.* films or compacted powders. It was assigned to the "softening point" T_s and the following collapse of the material inside the sample pan. It was therefore not due to a true thermal transition at all, but to a change in the physical state of the sample during heating, which caused a change in its thermal conductivity, *i.e.* a DSC instrumental artefact that, at first sight, had the appearance of a glass transition.[120]

DSC instruments can also be operated in the isothermal mode; this produces kinetic information, *e.g.* crystallisation or reaction rates. Note, however, that for such studies an isothermal calorimeter is to be preferred, because its sensitivity exceeds that of the typical DSC instrument by a factor of 10^4. Such methods will be briefly discussed in Section 11.6.

In recent times, various modifications of the basic DSC method have been reported. They include the so-called modulated DSC, in which a sinusoidally changing amplitude that is governed, as in standard DSC, by the temperature measured at the sample position.[140] The modification resembles in several aspects the features of Fourier transform spectroscopy. Its main advantage over standard DSC resides in its ability to differentiate between reversible and irreversible thermal effects.[174]

11.4 Fragility and Its Estimation

An estimate of the fragility of a given glassy preparation is required to enable a reliable calculation of its shelf life to be made. The estimation of the fragility parameter itself requires a number of thermoanalytical measurements to be performed. Three methods, in decreasing order of experimental rigour, are given below.

† They had been given various names by different investigators, *e.g.* "incipient melting" and "antemelting".[42] This subject has been raised at several international conferences and has led to heated debates; see for instance "Biological Product Freeze-Drying and Formulation, Dev. Biol. Standardization, Vol. 74, 1992, pp. 119–122.

11.4.1 Extrapolation of Configurational Entropy to Zero

The entropy change on crystal-melting, ΔS_m, may be taken to represent S_c at the melting point T_m, hence Equation (7) may be written as

$$\Delta S_m = \int_{T}^{T_m} (C_p^{conf}/T)\mathrm{d}T$$

For many organic glass formers, $C_p^{conf} = (\text{constant}/T)$ and C_p^{conf} can therefore be calculated at T_g (or any single temperature), permitting T_K to be estimated.

$$\frac{1}{T_K} = \frac{1}{T_m}\left(1 + \frac{\Delta H_m}{C_p^{conf} T_g}\right) \tag{9}$$

where ΔH_m is the enthalpy of fusion. ΔC_p, as determined by DSC, is not C_p^{conf} but is the difference in C_p between the undercooled liquid and the glass. Hence, absolute C_p measurements are needed for both crystalline and undercooled states, in order to calculate C_p^{conf} (but note that $\Delta C_p/C_p^{conf}$ is approximately 0.9). Assuming that $T_K = T_o$, and at T_g, $\tau \approx 10^2$ s (at a scanning rate of 10 K min^{-1}) and $\tau_o \approx 10^{-14}$ s (the time scale of vibrational motions), the VTF Equation (5) may be solved for *D*.

11.4.2 DSC Scanning Rate Dependence of T_g

T_g is the temperature *region* where an undercooled liquid deviates from, or returns to equilibrium, and it describes the relationship between relaxation time and temperature. The measured T_g depends upon the applied experimental heating (or cooling) rate, as described by Moynihan *et al.*:[141]

$$\frac{-\Delta E_{T_g}}{R} = \frac{\mathrm{d}(\ln q)}{\mathrm{d}\left(\frac{1}{T_g}\right)} \tag{10}$$

where q is the heating (cooling) rate, R is the gas constant and ΔE_{T_g} is the activation enthalpy of structural relaxation at T_g. Note however, that the non-Arrhenius behaviour of fragile materials close to T_g signifies that ΔE_{T_g} is large and therefore produces a smaller influence of q on T_g.

This can lead to significant experimental errors unless careful calibration is carried out throughout the course of experimental runs.

The temperature dependence of the apparent activation energy (ΔE^*) across a wide temperature range is given by

$$\frac{\Delta E^*}{R} \equiv \frac{\mathrm{d}(\ln\ \tau)}{D(1/T)} = \frac{DT_o}{(1 - T_o/T)^2} \tag{11}$$

Replacing ΔE^* with ΔE_{T_g} at a single temperature, an alternative fragility parameter m can be defined by

$$m \equiv \left.\frac{\mathrm{d}\ \log\ \tau}{\mathrm{d}(T_g/T)}\right|_{T=T_g} = \frac{\Delta E_{T_g}}{(\ln\ 10)RT_g} \tag{12}$$

which is sometimes written as

$$m = \frac{\Delta H}{(2.303RT_g)} \tag{13}$$

The parameter m can be linked to the VTF equation by Equation (11), viz.

$$m = \frac{DT_o/T_g}{(\ln\ 10)(1 - T_o/T_g)^2} \tag{14}$$

A high value of m indicates a rapidly changing mobility at T_g, *i.e.* fragile behaviour. The minimum value of m is given by

$$m_{\min} = \log\left(\frac{\tau(T_g)}{\tau_o}\right) = \left(\frac{D(T_o/T_g)}{(\ln\ 10)(1 - T_o/T_g)}\right) \tag{15}$$

Assuming again that at T_g, $\tau \approx 10^2$ s (at a scanning rate of 10 K min^{-1}) and $\tau_o \approx 10^{-14}$ s, the minimum value of m is found to be 16.

By combining Equations (14) and (15), D can be calculated in terms of m by

$$D = \frac{(\ln\ 10)m_{\min}^2}{m - m_{\min}} \tag{16}$$

A simpler graphical method makes use of Equation (10).[142]

Plotting ln q versus the reciprocal glass transition temperature allows a direct estimation of the apparent activation energy, which can be used directly in Equation (12) or (13), to evaluate the parameter *m*:

$$m = \frac{\Delta E_{T_g}}{2.303\ RT_g} \tag{17}$$

11.4.3 Glass Transition Width

An undercooled liquid has a distribution of relaxation times and hence the observed glass transition occurs over a range of temperatures. "Strong" glass formers do not show rapid changes in τ with temperature and therefore give rise to broad glass transitions. Fragile behaviour leads to a narrower glass transition temperature range.

Moynihan *et al.*[143] proposed the useful relations:

$$\left(\frac{\Delta E_\eta}{R}\right)\left(\frac{1}{T_g^{on}} - \frac{1}{T_g^{off}}\right) = \text{constant} \tag{18}$$

or

$$\Delta E_\eta = CR \bigg/ \left(\frac{1}{T_g^{on}} - \frac{1}{T_g^{off}}\right) = CRT_g^{on}T_g^{off}/\Delta T_g \tag{19}$$

as a means for the estimation of fragility. The superscripts "on" and "off" refer to the start and completion of the glass transition range as observed by DSC. In Equation (19), C is a constant and ΔT_g is the width of the glass transition. Experimentally, the proportionality constant $C = 5\pm0.5$.

Assuming an equivalence of ΔE_η and ΔE_{T_g}, then both m (pragmatically) and D (more rigorously) can be calculated, as discussed in Section 11.4.2.

In summary, all the three methods give similar results, but the calculation of m or D from the temperature dependence of the glass transition *width* is the most practical first-choice method, especially as the method that is based on the temperature dependence of T_g requires rather careful calibration of the instrument. Generally for pharmaceutical materials, T_g/T_m falls within the range 0.71–0.86 and D in the range 7–13, *i.e.* they are moderately fragile materials. There appears to be no simple relationship between T_g/T_m and fragility.

Finally, the question of relaxation below T_g and its relation to fragility requires comment. Experimental data suggest that pharmaceuticals are moderately fragile above T_g, but there seems to be no *a priori* reason why the same behaviour should be expected below T_g. The answer to this question is, however, important for the reliable prediction of shelf life. Apparent activation energies ΔE^* for moderately fragile pharmaceuticals have been found to be 200±40 kJ mol^{-1}. It is likely that moderately fragile materials would have a large distribution of relaxation times in the vicinity of T_g (see Equation (3) in Chapter 6). In practice, although there are some deviations from Arrhenius behaviour, the range of relaxation times is relatively constant and Arrhenius-like. Correlating molecular mobility (via τ) with physical and/or chemical stability may, however, not be a trivial exercise.

11.5 Monitoring Physical and Chemical Stability

For obvious reasons, the definitive test of stability must be the storage of the dried product at a specified temperature and performing periodic activity assays over the period of interest. Depending on the desired shelf life of products that exhibit low rates of deterioration, this procedure is time consuming, especially where the assay methods might be laborious to perform or require repeated calibration of expensive instrumentation, and/or highly skilled personnel. Conventionally, therefore, shelf lives of dried products are determined by performing a series of accelerated stability tests that involve stressing the product at temperatures above T_g, *i.e.* $(T - T_g) > 0$, and extrapolating the results to the desired temperature. This procedure yields unambiguous information, provided that any deterioration processes that occur exhibit Arrhenius kinetics, *i.e.* strong fluids. For fragile fluids, on the other hand, deterioration rates follow the VTF equation, *e.g.* Equation (4), or the Williams–Landel–Ferry (WLF) equation:[144]

$$\eta = \eta_o \exp\{C_1(T - T_g)/[C_2+(T - T_g)]\} \tag{20}$$

where $C_1 = DT_o/(T - T_g)$ and $C_2 = (T - T_g)$. The WLF equation is thus seen to be a special case of the VTF equation. It has been, and still is, used extensively in efforts to relate the textural properties of food materials and processed foods to the elusive property, termed "quality".[167]

As discussed in Section 11.2, physical deterioration, such as structural collapse, is not completely halted at $(T - T_g) < 0$, although such processes are markedly slowed down. Depending on the magnitude of $(T - T_g)$, relaxation rates may become too low for their measurement in real time. It is possible, therefore, that the measured deterioration processes may display changes in kinetic mechanisms with increasing temperature.

11.6 Estimation of Useful Shelf Life

Almost all chemical and physical rate processes or phase transitions are accompanied by energy changes. Thermoanalytical methods are therefore potent tools for monitoring such events, although they cannot in themselves provide any *direct* information of events at the molecular level. Three main groups of calorimetric methods find application: isothermal, adiabatic and temperature scanning. Under the right conditions, scanning methods are economical and convenient as monitors of overall long-term stability, and thus they find extensive application in industry.[145]

Accelerated chemical stability assays are a popular method for determining shelf lives of pharmaceutical preparations. A major advantage is their acceptance by regulatory authorities as a valid method for the determination of pharmaceutical stability. Samples are exposed to a range of temperatures above the desired storage temperature, and the stability is measured over a given length of time, in terms of the retained biological or biochemical activity. A typical protocol may include temperatures of −20, +4, +20, +37 and +45°C and storage periods of 3 and 6 months, followed by 1, 2 and 5 years. Results are usually extrapolated back to the recommended storage temperature, frequently 25°C. An undoubted advantage is the fact that the assay can be made to be specific for a given chemical change. On the other hand, the uncertain extrapolation of data, taken at high temperatures, to 25°C is a distinct disadvantage and can lead to considerable uncertainties. The methods are also time consuming and assays may well be laborious to perform. It has been shown that even when the loss of a given protein activity appears to exhibit Arrhenius kinetics, this may hide a combination of individual degradation mechanisms, although giving rise to an apparently simple overall inactivation profile.[176]

By contrast, DSC provides a fast and economical alternative. However, because of the marginal sensitivities of some commercial instruments, high temperatures and long extrapolations sometimes become necessary. The method is also non-specific, in the sense that energy changes, rather than chemical changes, are measured.

Quantities directly measurable by DSC are shown in Figure 9; they include:

- glass transitions and their time dependence, $T_g(t)$;
- softening points, T_s, the temperature at which the viscosity (mechanical relaxation) becomes measurable in real time ($\eta \cong 10^8$ Pa s), indicated in Figure 5;

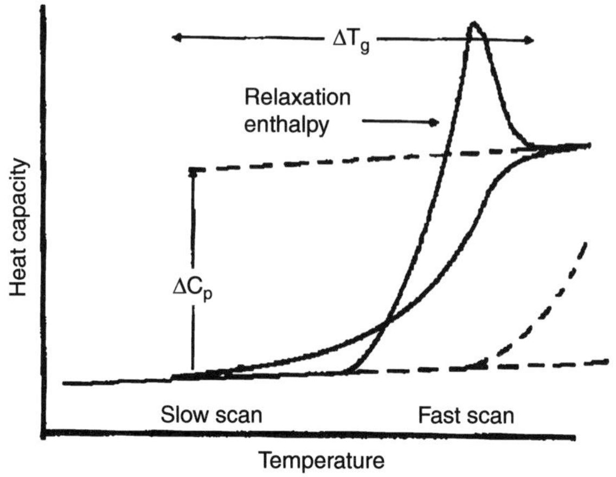

Figure 9 *Diagrammatic representations of typical DSC scans of the glass transition showing the quantities required for the determination of the fragility parameter and T_o. The first scan, usually taken after storage (annealing), reveals an endotherm of the relaxation enthalpy, which disappears after recooling and immediate reheating*

- width of the glass transition, ΔT_g;
- relaxation enthalpy over time, $H(t)$;
- absolute specific heat over time, $C(t)$;
- specific heat of the glass transition, ΔC_g;
- cooling/heating rates, q_c and q_h.

With the aid of these experimental quantities, the following parameters can be calculated:

- relaxation time and its temperature dependence, $\tau(T)$;
- energy of activation of the relaxation process, ΔE^*;
- Kauzmann temperature (temperature of zero mobility, T_o);
- fragility parameter, m;
- temperature dependence of viscosity, $\eta(T)$.

Isothermal microcalorimetry has the distinct advantages of convenience, speed and high sensitivity. Moreover, determinations are performed at 25°C or whatever might be the desired storage temperature. There is thus no need to perform uncertain extrapolations of results obtained over a range of high temperatures. The basis of the technique is given by the following relationship, which forms the connection between thermochemistry and kinetics:

$$(\mathrm{d}q/\mathrm{d}T) = (\mathrm{d}n/\mathrm{d}T)\Delta H \tag{21}$$

where the left-hand side is the heat flow per second, the first term on the right-hand side is the number of mols reacting per second, and ΔH is the enthalpy of the reaction.

For a general reaction, such as aA + bB → cC, the heat flow can be expressed as

$$(\mathrm{dq}/\mathrm{dt}) = k\,\Delta H[A - (aq/\Delta H)]^m[B - (aq/\Delta H)]^n$$

By plotting power (dq/dt) against heat flow (q), the constants ΔH, k, m, n, A, B can be determined by a process of iteration.

For processes with $\Delta H \cong 40$ kJ mol^{-1}, microcalorimeters can detect reaction rates corresponding to 6 μmol month^{-1}. Alternatively, for first-order rate constants of 10^{-10} s^{-1}, half-lives of 200 years can be conveniently measured (equivalent to reaction rates of 0.3% per year). For examples of various aspects of isothermal microcalorimetry, see Refs. 146–148.

As discussed for DSC earlier, disadvantages of microcalorimetry include its chemical non-specificity and also a possibility that unpredictable physical changes might lead to non-Arrhenius behaviour. Such changes include unexpected phase transformations, water transfer between excipients and drugs, improperly defined rate constants, and parallel reactions of the same order but with different activation energies.

11.7 Dynamics in Supersaturated Solutions

Before discussing the role of any residual water that remains within the solid solution after freeze-drying, we shall consider the dynamics of water and other molecules within supersaturated solutions, such as are formed in aqueous PHC solutions during any drying process. A ground-breaking study has revealed that, although completely miscible on a macroscopic scale, extreme complexities arise in the diffusional motions of water, relative to those of the PHC molecules, as their concentration increases beyond the saturation value.[149] We report here the major results and conclusions of this study because of their high relevance to an understanding of the molecular-scale processes that accompany the concentration and temperature changes during drying.[‡]

The study encompassed measurements of rotational and translational diffusion rates of different parts of PHC molecules (by ^{13}C NMR) and –OD, –CH_2OD and D_2O "hydration phase" (by ^{2}H NMR) over a range

[‡]Some of the contents of the thesis have since been published in English.[150,151,156] Although the main experimental details and data are recorded in the published papers, their discussion in terms of the molecular dynamic details is hardly touched upon; they can only be found in the thesis itself. The translation and condensation to a 15-page English summary has been performed by F. Franks and can be obtained from him at asdi35@dsl.pipex.com.

of temperatures (220–350 K) and concentrations (10–70% by weight) for sucrose, α,α-trehalose, maltose and α-D-methyl glucoside. The following three questions were addressed:

(i) How do sugars affect the dynamics of water?
(ii) How do the dynamics of sugar molecules depend on the concentration?
(iii) What conditions determine the "transition" into the vitreous state?

The ^{13}C relaxation results led to the following conclusions:

- Within the concentration (c)/temperature domain studied, sugar molecules move as rigid bodies.
- For $c < 30\%$, the temperature dependence of the rotational correlation time τ follows the Arrhenius law.
- At $c > 30\%$, the temperature dependence of τ follows the VTF law, Equation (5). The dependence of τ on c is determined exclusively by $T_o(c)$.
- For temperatures below $T_o(c)+135$ K, deviations from isotropic tumbling become significant.
- The temperature of zero mobility, T_o, correlates well with the calorimetric glass temperature T_g, where $T_g=(T_o+20\text{ K})$.
- The $-CH_2OH$ groups display an additional, hindered rotation about the symmetry axis. For $c > 30\%$, the ratio τ_{ring}/τ_{CH2OH} increases strongly.

Results of ^2H relaxation measurements at elevated pressures:

- Up to $c=30\%$, the isothermal pressure dependence of rotational motion is anomalous; an increase in pressure produces an increased water mobility.
- At $c > 50\%$, the pressure dependence disappears.
- The temperature dependence of τ_{water} follows the VTF law.
- A hydration model, in which nearest water neighbours are considered to form a quasi-phase, suggests that "hydration water" performs anisotropic rotation.

Results of ^2H relaxation measurements at ambient pressures:

- Up to $c = 40\%$, rotational motions of water decrease rapidly with decreasing temperature.

- The isotropic reorientation of the "hydrated complex" has the same τ value as that found for sugars by ^{13}C relaxation. Below 270 K, no dynamic equilibrium between bulk and hydrate water is detected; the correlation times diverge and T_o values are no longer identical.
- At $c > 60\%$, the mobility of sugar and hydrate water freezes out at $T < 200$ K. In the water phase, free rotation persists down to $T \approx 120$ K! The calorimetric T_g values correspond solely to those of the PHC matrices.

In summary, the results can be treated in terms of three distinct concentration domains:

(i) $c \leq 30\%$: PHC molecules perform uncorrelated motions, and the solution viscosity exhibits a minimal concentration dependence. The dynamics of water molecules are governed by co-operative fluctuations within the hydrogen-bonded water network. In undercooled solutions, the network increasingly inhibits the orientations and the rotational correlation time increases steeply with decreasing temperature. Hydrostatic pressure distorts the network and, in the metastable phase, it inhibits the generation of ordered domains with linear H-bonds; molecular motions thus remain rapid. Dissolved PHC molecules, because of their incompatible –OH orientations, produce a similar effect. This makes it possible to undercool PHC solutions far into the metastable region, resulting in eventual vitrification.

(ii) $c > 40\%$: Hydrated PHC molecules form aggregates that develop into a network. The resulting change in the short-range structure leads to a distribution of correlation times. The water molecules are integrated into the network, and their rotational and translational motions are affected. This is strikingly demonstrated in Figure 10, which shows the ratio of sugar:water τ values as a function of concentration. An increasing coupling of diffusive motions (microheterogeneity) between sugar and water is evident in a complex mixture; it reaches its maximum value at $c = 50\%$. At higher concentrations, PHC molecules form networks via direct H–bond links, with the expelled water molecules collecting in microscopic droplets; their mobility is facilitated in this quasi-heterogeneous system. At this stage, PHC and water motions lose their coupled motions, with a sharp rise in the τ ratio.

(iii) $c > 70\%$: A three-dimensional PHC network is formed. With decreasing temperature, the network forms a macroscopic gel and the PHC molecules are unable to adopt the crystal configurations

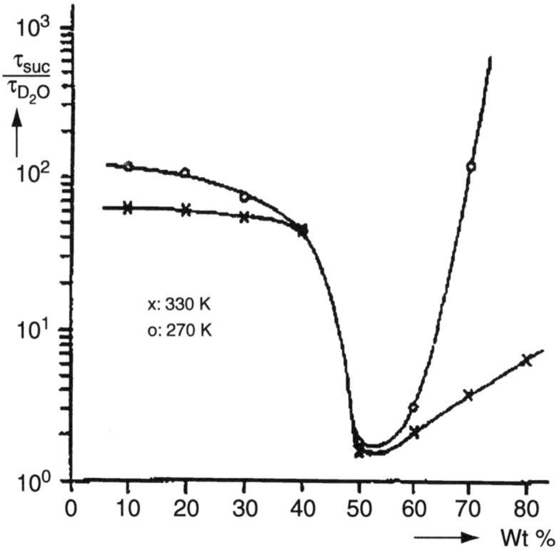

Figure 10 *Nuclear magnetic relaxation times of sucrose relative to that of water as a function of sucrose concentration and showing the phenomenon of maximum dynamic coupling between sucrose and water at 50% w/w (4 mol%), which disappears, as the glass transition is approached. Redrawn from Girlich and Lüdemann*[156]

in real time. The metastable mixture forms a glass. Water molecules are trapped in the free volume within the glass; their translational diffusion is (almost) blocked, *i.e.* they become osmotically inactive, although their rotational freedom persists down to 120 K.

None of these dynamic effects are revealed in the macroscopic thermodynamic properties of PHC–water mixtures, nor can they be studied by X-ray diffraction because of the short range of the microheterogeneities. In principle, neutron scattering might provide more information, but at the time of writing, no reports of such studies have yet found their way into the public domain. However, the comprehensive study by Girlich[149] of such mixtures, involving multinuclear relaxation measurements over extended pressure–temperature–composition phase domains, is of a quality rarely found in the scientific literature; it has revealed much detailed information that is of relevance to ecology, climate physics and several branches of technology, freeze-drying being among them.

11.8 Residual Water and Its Estimation

At the completion of primary drying, more than 99% of the original water in the formulated solution would have frozen; the remaining water content of an amorphous product is generally expressed as W'_g g per g solid. The actual value, as shown in Figure 9 in Chapter 5, typically 0.2–0.3 g g^{-1}, is partly determined by the freezing behaviour of the formulation during cooling. It is likely to vary from product to product or even from vial to vial. DSC is commonly employed to obtain an estimate of this residual water, which forms part of the glassy phase, and that must be wholly or partly removed by secondary drying.

A "correct" definition of "dry" has already been commented upon because even the most hydrophobic organic substances, *e.g.* polystyrene, absorb water vapour to some extent. Water has therefore been called Nature's "ubiquitous plasticiser". Apart from this softening role, water, being itself a very reactive and versatile molecule, might also have other effects on amorphous solid substrates, some of them unexpected, and most of them deleterious. The role of low water contents (< 1% w/w) in affecting the behaviour of solid solutions is a subject that deserves further investigation.

The reliable estimation of residual water in dried solids is of importance but is beset by several problems,[152] mainly related to the shape and interpretation of DSC heating traces, as illustrated in Figure 11 for a typical aqueous mixture, maximally frozen, from which any relaxation enthalpy contribution has been removed by annealing.[117] The drawn-out DSC heating trace represents the superposition of several distinct processes: T'_g of the mixture, the heat of dilution, produced by ice

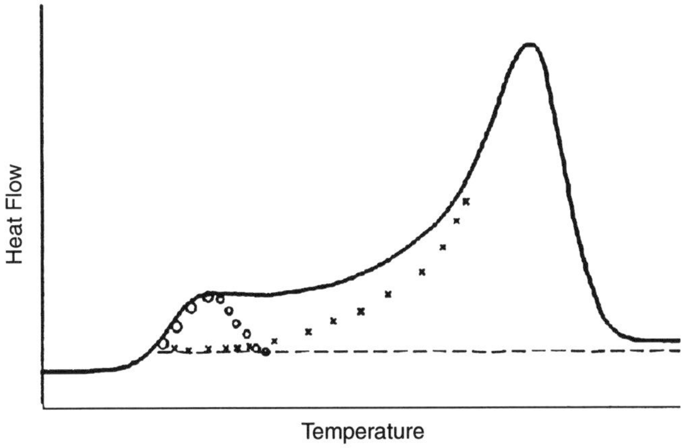

Figure 11 *Contributions (not drawn to scale) to the total measured glass transition and ice-melting endotherm for a frozen aqueous solution: crosses represent contribution from the heat of dilution of the supersaturated solution, circles denote the enthalpic relaxation contribution and the broken line is the glass transition. The drawn out curve represents the measured overall heat flow, and the area under that scan corresponds to the total enthalpy change.*

melting back into the highly supersaturated mixture, the latent heat of fusion of ice, and the difference in heat capacities between ice (+ solid solution) and liquid solution, after all the ice has melted. This latter contribution is represented by the baseline shift between the glass and the liquid. The practical problem lies in deciding on the correct assignment of baselines and the deconvolution method used to arrive at the contribution due solely to ice melting, because it is the only part of the area of the total endotherm that is related to the quantity of unfrozen (residual) water remaining in the freeze-concentrated mixture. Even if this operation can be performed with some degree of confidence, which is doubtful, there remain other problems, *e.g.* the correct assessment of the latent heat of fusion of ice, dispersed as microscopic crystals in a supersaturated aqueous mixture, at some subzero temperature.[171]

Turning now to the estimation of water in the dried preparation, one of the various approved modifications of the classical Karl Fischer technique is usually employed. This topic has been expertly treated in several monographs.[153,154] Great care must be taken to keep all the reagents dry and to prevent the pick-up of water vapour during the actual estimation. One common procedure involves the estimation of water at various times after breaking the vial seal and extrapolating back to zero water content. The weakness of such a method, in common with similar methods relying on back extrapolation, lies in the possibility that there may occur a physical and/or chemical transition that alters the distribution of water within the product, *e.g.* the crystallisation of a hydrate or a change in the crystalline: amorphous ratio.

A weakness, common to all Karl Fischer-type methods, lies in the limitation that they measure the *total* water content of the sample, irrespective of the water distribution within the sample. In solids that are partially crystalline and partially amorphous, the residual water will be concentrated in the amorphous phase, thus depressing its T_g. This can accelerate or even promote the crystallisation of small molecule substances within the amorphous matrix.[146] Take as an example crystalline sucrose that contains 0.5% of amorphous material and 0.17% of residual water. Since all the water is concentrated in the amorphous phase, the "real" water content will be 20% with a T_g of 9°C. It is also instructive to calculate the number of water molecule layers for differently sized sucrose particles. This is shown in Table 1. If the measured water content were to rise to 0.5%, corresponding to 50% in the amorphous phase, then T_g of the amorphous phase would be depressed to −70°C. It is therefore useful, if not essential, to have a reasonable estimate of the amorphous content of a preparation. Several more or less laborious methods for its determination find application, and they are

Table 1 *Water layers surrounding spherical sucrose particles at 0.1% residual water content*

Water molecular layers (μm)	*Surface area (cm^2 g^{-1})*	*Water layers*
1	38,000	1.1
10	3800	11
38	1000	42
100	380	110

described in the following paragraphs in increasing order of the necessary experimental effort.

From a knowledge of the crystal-melting temperature, T_m, the glass temperature can be estimated with some degree of confidence, at least for many of the excipients commonly used for freeze-dried pharmaceutical preparations, where it has been found empirically that $T_g/T_m \approx 0.73$. Furthermore, for the same preparations, the temperature of zero mobility can be approximated by $T_o \approx T_m/2$. Also, for many materials, $\Delta C_g T_g \approx$ 100 kJ g^{-1}. At T_g, the viscosity of most glasses is assumed to lie between 10^{11} and 10^{13} Pa s, and above T_g, the viscosity often follows the WLF equation. It is thus possible to obtain T_o at ($T > T_g$). As shown previously, the fragility can be obtained from carefully performed DSC heating scans, with the aid of measured values for ΔC_g, T_g/T_m and ΔT_g.

From the ideal mixing rules and the Gordon–Taylor equation,[95] an estimate can be obtained for the effect of sorbed water on T_g. It is thus possible to determine the critical plasticiser content at which T_g corresponds to the ambient temperature. A fully amorphous sample can then be prepared by quenching a fused preparation into a cryogen, followed by freeze-drying or spray drying to complete dryness. The thermophysical characteristics of the fully amorphous, anhydrous substance can then be obtained. Water vapour sorption measurements should then be performed at several temperatures. The recrystallisation characteristics above T_g in the anhydrous state and in the presence of water vapour at a range of relative humidities can now be determined. Powder X-ray diffraction studies will indicate whether the substance crystallises as a hydrate and/or a metastable polymorph, such as might introduce mechanical/chemical instability. In order to estimate stability below T_g, it is necessary to determine enthalpic (also structural and/or chemical) relaxation rates under different conditions.

A direct experimental method that does not rely on approximations and assumptions has been reported by Ahmed *et al.*[147] It is based on a combination of DSC with isothermal calorimetry and powder X-ray diffraction. It is particularly useful for the detection of low amounts of amorphous material in allegedly crystalline drug products.

11.9 Residual Water: Its Influence on Stability

Opinions differ as to the optimum residual water content to be aimed at during the secondary drying stage. Those "skilled in the art" generally recommend levels in the range 0.5–3.0% of the dry substance. On the other hand, it has often been reported that "overdrying" reduces the stability of some products.[115] Various reasons are advanced, *e.g.* the removal of water from native protein structures can result in their collapse and irreversible inactivation. It is the opinion of the author that the symptoms ascribed to "overdrying" are produced by faulty processing conditions during the secondary drying stage, such that for reason of the economics of speed, the product temperature is raised too far above the softening point, which usually lies close to T_g+10 °C; this is a recipe for disaster (see Figure 4 in Chapter 9).

11.10 Unresolved Issues

Although considerable progress has been made in the elucidation of relaxation mechanisms of pharmaceutical glasses, there still exist several unresolved questions, all of which have a bearing on the reliability of accelerated storage assays. They include:

(i) Is there a relationship between T_g and chemical stability? In other words, are mechanical and chemical deterioration processes related? Few reports exist on chemical changes in glasses. Figure 12 shows WLF plots of hydrolysis rates of 2–(4-nitrophenoxy)tetrahydropyran in a variety of PHC glasses.[155] Three features are of interest:

- All the ln k/T curves are non-monotonic, with inflexion points at T_g.
- The curves suggest that well below and well above T_g the hydrolysis reaction follows Arrhenius kinetics, but in the neighbourhood of T_g deviations are observed.
- The curves indicate that substances with *low* T_g values (sucrose) are superior in retarding the hydrolysis reaction, even though they do not provide the same degree of thermomechanical stabilisation as PHCs with higher T_g values (dextran).

(ii) At temperatures at and below T_g, is the long-term chemical stability of a product related to the structural relaxation time? If so, can the chemical stability be calculated from experimental measurements in real time? *Physical* stability is clearly related to

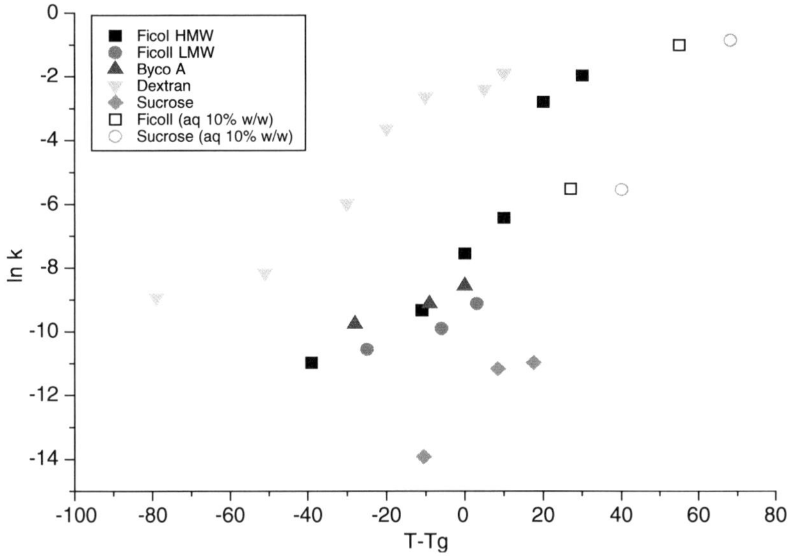

Figure 12 *WLF plot for the hydrolysis kinetics of 2-(4-nitrophenoxy)tetrahydropyran, embedded in matrices of PHC glass formers in the neighbourhood of T_g. Note that the degrees of stabilisation provided by the glasses are in the reverse order as the glass transitions of the PHC; i.e., dextran provides the largest (ln k) decrease, but sucrose provides the most pronounced hydrolysis inhibition*

viscosity, and hence, to structural relaxation. *Chemical* stability may be subject to additional factors, *e.g.* possibility of reactions triggered by small molecules (water, oxygen and ammonia) or intramolecular rearrangements. Investigations of chemical deterioration reactions in different biopharmaceutical formulations demonstrate that reaction rates are not correlated with T_g values of the formulated preparations. On the other hand, as shown in Figure 13, very definite correlations exist between rate constants of some typical chemical reactions undergone by freeze-dried human growth hormone (hGH) preparations and T_o values of the formulated solutions.

(iii) How are intramolecular processes affected by vitrification, *e.g.* deamidation or racemisation of proteins?

(iv) What, if any, is the effect of residual water on the long-term stability, apart from its depressant action on T_g?

(v) What is the relationship between molecular structure, size, flexibility and T_g (and T_o)? Many examples exist of pairs or larger groups of stereoisomers (sorbitol/mannitol; α-trehalose/β-trehalose; glucose/mannose/galactose/fructose) with significantly different glass temperatures.

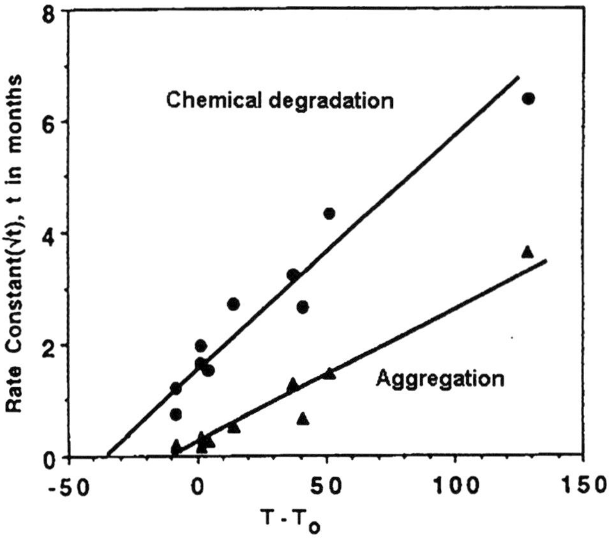

Figure 13 *Relationship between kinetics of chemical processes, measured as $t^{1/2}$ versus $(T - T_o)$ of freeze-dried hGH, formulated with stachyose and trehalose, with protein:sugar weight ratios of 1:1, 3:1 and 6:1, at 40 and 50°C. Symbols: • = chemical deterioration and ▲ = hGH aggregation. Redrawn with changes from M.J. Pikal, private communication of unpublished results*

Possibly several of these remaining puzzles are interrelated. Until cause-and-effect relationships can be established, there remain uncertainties about predictions of long-term behaviour of freeze-dried preparations, based solely on physical measurements in real time. For safety considerations alone, therefore, reliable predictions will presumably have to involve laborious and time-consuming real-time chemical assays.

11.11 Reconstitution at Point of Administration

The delivery of freeze-dried preparations can be performed by different routes: oral, nasal, anal, pulmonary, transdermal and parenteral. Of these routes, some do not require any treatment of the drug before it is administered, *e.g.* in the form of powders or tablets or in inhaling devices. For parenteral administration, however, whether by injection or infusion, the freeze-dried cake must be returned to a liquid state, a process referred to as "reconstitution". The main vehicle will normally consist of "water for injection" or a solution, the concentration of which will establish isotonicity. The time required for the complete dissolution of the cake may in some cases be critical and should therefore be known.

CHAPTER 12

Case Histories

12.1 Introduction

Successful freeze-drying, on a commercial scale, of essentially labile materials depends on a detailed understanding of the complex interplay between several disciplines: ice physics, chemistry of supersaturated solutions, pharmacology, materials science of water sensitive materials, engineering of coupled heat and mass transfer, production economics and IP law. The case histories discussed in this chapter originate mainly from projects undertaken by the author and his colleagues at Pafra Biopreservation during the period 1985–1999 and from various consultancy projects undertaken by the author since that time.[†]

A logical sequence of activities is recommended for the co-development of a specified product and an optimised process cycle. During the pre-formulation and formulation stages, it is well to carry out a literature search in order to ensure that the actual formulation arrived at does not run the danger of excessive IP litigation. Thus, a large international company was in the process of developing a new product, aiming to use alanine as the stabilising excipient. The only reason for this choice turned out to be that a competitor had patented the use of glycine for a similar product. It was nevertheless intended to use the same drying cycle as the one described by the competitor, a sure recipe for disaster.

After a thorough literature search, the suggested sequence to be followed begins with pharmacokinetics. Assuming that a specified unit dose of drug is to be delivered, it must be established that none of the other components in the formulation interferes with the action or the potency of the drug substance. The formulation is then built up, taking

[†] Where the results have not been published in the public domain, names of client companies are not divulged, although the period of confidentiality has expired in every case.

into consideration pH, isotonicity, total solids composition and concentration, vial fill volume and the desired thermomechanical properties of the dried cake (here a suitable T_g value is a first approximation). Vial geometry needs to be considered because it will determine the solution fill depth and also the loading of the freeze-drier. Then follows a freeze–thaw stability trial, which must be successful before any freeze-drying can be contemplated. Finally, and of the utmost importance, T'_g of the chosen formulation must be determined. If this step is omitted, then freeze-drying becomes purely a trial-and-error process.

In tandem with the formulation development, certain process parameters pertaining to the drier should be known. These will usually have been established during the initial validation of the equipment. It is useful to run a trial with a fully loaded drier, using the same vials that will be used later, but filled with water to 1 cm depth, loosely stoppered. The loss of water in a number of vials at certain time intervals is then monitored. The results will provide useful information about the heat transfer coefficient K_v (see Chapter 8) and also about the uniformity of the sublimation rate across and between shelves (detect possible hot spots and cold spots). The information will also prove useful when scale-up of the pilot process requires certain changes to be made in the process parameters.

Having estimated the optimum primary drying parameters, a pilot study should be performed. It is not good practice to carry out such a study either in an incompletely filled drier or by filling up shelves partly with empty vials. The reliability of pressure and temperature measurements should be checked at this stage. It has sometimes been found that the (apparent) product temperature exceeds the shelf temperature. Since this is an obvious physical impossibility, such an anomaly is due to faulty calibration of the thermocouple probe.

Secondary drying is preferably carried out by using the temperature ramping method, described in Chapter 9. It must first be ascertained that the "rest" period between heating steps is long enough to achieve thermal equilibrium in all vials and that the temperature steps are not unduly large (see Figure 5 in Chapter 9). The cycle can be terminated when the desired glass temperature and residual water content have been reached.

At the completion of the cycle, selected vials are subjected to stability trials, which can take different forms. Accelerated (thermally stressed) trials are rapid, but subject to various types of uncertainties, and not as informative as isothermal procedures. Very long-term trials will establish whether water vapour, however generated, has entered a vial and damaged the product.

12.2 Choice of Formulation

A detailed account, describing the stepwise development of a freeze-drying process for a protein drug, has been published previously by the author's laboratory.[157] The following account is based on the cited report but is limited only to its salient points. The factors that must be considered during an initial formulation stage have already been discussed in other chapters. Pharmacokinetic aspects apart, they include the important decision whether to aim for physical homogeneity or heterogeneity, expressed by amorphous or crystalline end products, respectively. For biopharmaceutical products, especially those based on proteins, this choice seldom arises because crystallisation is generally not an option. For parenteral products, important questions concern the isotonicity and the volume of reconstituted solution to be administered.

12.3 Process Cycle Optimisation

Optimisation in this context usually means the shortest cycle commensurate with an end product of the desired quality. As discussed before, the only process variables that permit direct control are the shelf temperature, the chamber pressure and the duration of the total process cycle. Other things being equal, process control by adjustment of the chamber pressure is to be preferred over control by shelf temperature. During the secondary drying stage, however, the temperature needs to be raised because the diffusion of water from the solid cake is hardly affected by pressure. The heating rate must however be carefully controlled, so that collapse is avoided. The completion of the process may be judged by the residual water content, which is of course an indirect measure of the glass transition temperature T_g.

12.4 Concurrent Product and Process Refinement

The development here described involved a 4-amino-methylthioxanthone derivative (WIN) with novel cytotoxic antineoplastic activity. Its properties had been evaluated in clinical studies in the form of a stabilised solution formulation, supplied in ampoules at a concentration of 2.5 mg ml^{-1} in citrate buffer (pH 5.5). Unless kept refrigerated (2–8°C), a precipitate of drug dimer was gradually formed during storage. The main aim of the freeze-drying optimisation process was to provide a commercially acceptable drug that could be stored at ambient temperature. Three trial formulations were prepared, as detailed in Table 1.

Table 1 *Aqueous formulations of WIN, before addition of stabilisers and their physical states after freeze-drying*

		Formulation		
Component	*Physical state*	*1*	*2*	*3*
WIN (mg)		10.0	10.0	10.0
Acetic acid (0.1 M, in ml)	Crystalline[a]	0.5		
Citric acid (mg)	Partially crystalline		9.6	
Lactic acid (0.1 M, in ml)	Vitreous			0.5
NaOH (0.5 M, in μl)			8.8	12
Water for injection (ml)		1	1	1
pH		4.95	3.98	3.99

[a] Volatile buffer, evaporates during drying.

Mannitol, dextran and sucrose were subsequently employed as stabilising excipients at 50 mg ml^{-1} levels.

Calorimetric analysis was performed by a modified differential scanning calorimetry (DSC) procedure.[158;‡] HPLC was used to analyse the reconstituted solutions. After freeze-drying, the DSC traces of the three formulations shown in Table 1 displayed the three states that are typically encountered in freeze-dried solutions, namely: complete crystallisation, partial crystallisation and complete amorphisation (glass formation).

Exhaustive reformulation, thermal cycling and annealing studies with addition of the excipients at several concentrations led to the following T'_g data: lactate and dextran containing solutions: −19°C; mannitol containing solutions: *ca.* −30°C. On reheating, mannitol invariably crystallised, and the phenomenon of vial cracking, already earlier reported,[159] was observed.

Reformulation results indicated that an improved preparation, coupled with a shorter drying cycle, could be achieved in lactate/sucrose stabilised systems, as follows: the drug concentration in the initial solution could be increased to 20 mg ml^{-1}, thus halving the fill volumes and, with an unchanged T'_g, a reduction in the primary drying time could be achieved. Isotonicity could be adjusted by the addition of NaCl. Examples of the "fine tuning" formulation experiments are shown in Table 2. Formulations A and B were favoured because of the higher temperature that could be employed for primary drying. Of the two, Formulation B gave a higher sublimation rate because of its lower solid content, whereas Formulation A, with the higher solid content, resulted

[‡] At that time, commercial software for the investigations of DSC recorder traces for signs of glass transitions were not yet available.

Table 2 *"Fine tuning" test formulations used to determine the relative effects on T'_g values of NaCl and sucrose on buffered WIN solutions*

Component	*Formulation* *A*	*B*	*C*	*D*
WIN (mg)	20	20	20	20
Lactic acid (0.1 M, in ml)	0.5	0.5	0.5	0.5
NaOH (0.5 M, in µl)	12	12	12	12
Sucrose (mg)	100	50	40	30
NaCl (mg)	2.7	2.7	3.8	5.0
Water for injection (ml)	1	1	1	1
pH	4.0	4.0	4.0	4.0
T'_g (°C)	−37	−38	−42.5	−45.5

in a mechanically stronger cake structure. A preference for Formulation B was decided upon, and test samples were freeze-dried (primary drying temperature: −40°C). The mean residual water content of optimally dried vials was 1.4% (Karl Fischer coulometric analysis), with $T_g = 42$°C, giving a maximum "safe" storage temperature of *ca.* 37°C. Dried samples were subjected to extended stability trials at 30 and 40°C. The product was shown to be chemically stable; no changes in assay or pH were observed after 6-month storage.

This discussion demonstrates how a logical refinement of acceptable chemical formulation parameters against measured thermomechanical properties of the dried preparations will aid in the development of stable products. Thus, the often-repeated dictum,[13] even by those who should know better, that acceptable products can only be obtained by trial-and-error methods is outdated and should be discarded.

12.5 Inadvertent Failure: "Unexpected Problems"

On occasions, even long-established processes may give rise to a rogue failed batch of vials. Failure through collapse of the cake will be easily observable, but that may not be the case when the failure resides in a partial or complete loss of biological activity of the product without observable changes in the ultrastructure of the cake. It then requires a "forensic" approach to investigate the causes of failure. This event happened with one isolated freeze-drying run of Plasma Coagulating Factor VIII, which had been freeze-dried on a large scale by a well-established process cycle. Comparative tests on successful and failed products revealed that failure was due to an unexpectedly low T'_g value of −55°C, which, according to specification, should have been −45°C.

Since solutions were routinely frozen to $-50^{\circ}C$, it was evident that the failed product had not been subjected to maximal freeze-concentration and the residual solution therefore contained an excess of unfrozen water, causing it to undergo collapse during primary drying. The results are shown as electron micrographs in Figure 1, and the corresponding DSC heating traces of the reconstituted solutions in Figure 2.[71]

It became clear that the cause of failure resided in an anomalous formulation of a particular batch of vials. The method of preparing the solutions relied on the addition of known volumes of buffer, excipients and protein solution, the latter taken straight from the chromatographical column, to the vial, but without assaying the protein content. The protein solution used in the preparation of the failed batch of vials was later found to have contained an abnormally low amount of protein that, in turn, gave rise to the depressed T'_g.

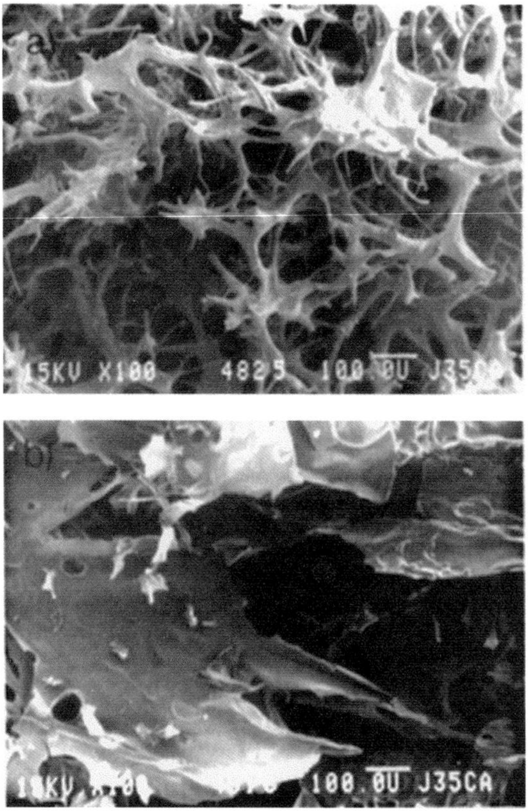

Figure 1 *Scanning electron micrographs of freeze-dried Factor VIII preparations, taken from: (a) a regular production batch with fine, porous structure; and (b) a failed batch showing extensive cake collapse; bar lines represent 100 μm*

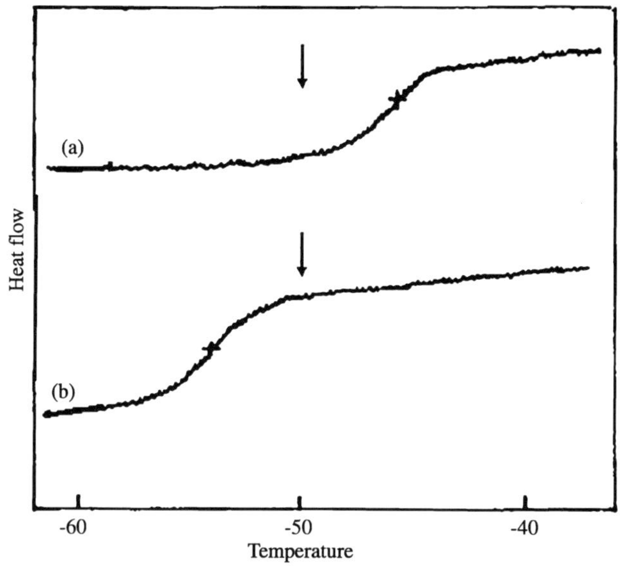

Figure 2 *DSC power–time scans of reconstituted Factor VIII solutions after freeze-drying: (a) successful product (see Figure 1a) and (b) failed product (see Figure 1b). Crosses mark T_g' and the arrows mark the temperature to which the solutions were frozen prior to the start of primary drying*

Other "unexpected" problems can arise from *in situ* changes in the formulation during storage. Such changes are usually due to some form of recrystallisation process, accompanied by the release of water. They are detected by their random occurrence and vial-to-vial variations in the residual water content within a production batch. A common form of *in situ* formulation change is observed with polymorphic transitions, particularly of hydrates. Such a problem occurred with a protein-based product that was stabilised with mannitol. The particular freeze-drying cycle, developed in the author's laboratory, had been devised to ensure that the mannitol did not crystallise during drying. A batch of test vials was delivered to the client, and the routine stability trials, undertaken at the client's QA facility, consisted of periodic tests for biological activity after storage at −20 and +37°C. No problems were detected for the vials stored at −20°C for 26 weeks. However, a vial from the 37°C storage lot exhibited a 20% activity loss after 2 weeks, and the next test, after 4 weeks of storage, gave a 75% loss. After 6 weeks of storage, the tested vial showed only a 20% loss, and this was repeated after a further 2 weeks. Thereafter (12 weeks of storage), a 75% loss was once again recorded. The test results are shown in Figure 3. A replacement batch of vials was dispatched to the client for further

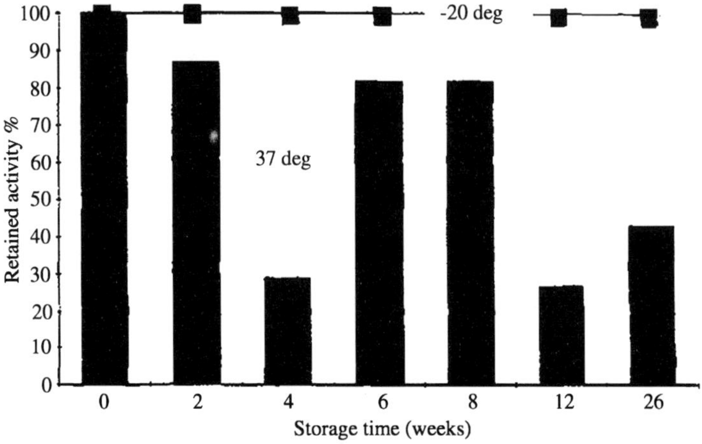

Figure 3 *Activity assays of a product, stabilised with mannitol, and stored for up to 26 weeks at −20 and +37°C. For details, see text*

stability tests. A set of random results, similar to that shown in Figure 3, was obtained.

It was pointed out to the client that during freeze-drying, great care had been exercised to ensure that the preparation, including the mannitol, remained in the amorphous state. On the other hand, stability trials carried out at 37°C ensured that mannitol would crystallise sooner or later, with the water remaining in the amorphous product phase, thus reducing its T_g and the stability of the product, but in an erratic manner, as was indeed observed. The curious stability data did not represent a recovery of activity in vials that had earlier given a poor result. Stability assays had been performed on vials, chosen randomly from a batch. Because of the random nature of mannitol crystal nucleation, any assay result depended entirely on the physical state of the dried product in a particular vial at the time of testing that, in turn, was governed by the absence or occurrence of mannitol crystallisation and the degree to which mannitol had already crystallised in the vial under test at the time. If stability at 37°C was essential, then a major reformulation was required.

12.6 Unfamiliarity with Freeze-Drying Parameters

A freeze-drying process was developed for a client for a preparation that contained 2 mg of a bioactive (cytotoxic) material and 51 mg of excipient, made up to 10 ml with water; process parameters are shown in Column 2 of Table 3. Starting with a measurement of T_g', and the

Table 3 *Calculated freeze-drying cycles for preparation P under different experimental conditions (Pafra Biopreservation, unpublished data)*

T'_g (°C)	−33	−33	−36
Unfrozen water content of freeze-concentrate (g g^{-1})	0.49	0.49	1.2
Vial diameter (mm)	46	26	26
Fill depth (mm)	6	18	18
Sublimation rate (g $vial^{-1}$ h^{-1})	1.47	0.27	0.17
Total mass of ice (g)	9.7	9.7	9.7
Primary drying time (h)	7	37	58
Secondary drying time (h)	12	16	18

given details of concentration, fill volume and vial dimensions, the cycle was calculated and verified experimentally. At some later stage, the client returned, stating that the cycle no longer worked and that total collapse was observed in all vials. After some discussions, it was finally established that a different set of narrower vials had been substituted, in order to increase the loading capacity of the freeze-drier. However, no allowance had been made for the consequences of an increased fill depth. A recalculation was performed, and the details are shown in Column 3 of Table 3. Thus, although the freeze-drier was now able to accommodate a larger number of vials, it was questionable whether the economies achieved balanced the necessity of a considerably longer primary drying period.

Even later, the same client returned with a further "complaint". A "minor" change in the formulation had been performed by a "small" increase in the content of bioactive material to 5 mg. Freeze-drying had failed to provide a successful product, despite the employment of the recommended cycle. Since a change had now occurred in the composition of the solution, a further T'_g measurement had to be performed, with the result of −36°C, thus requiring a complete recalculation of the drying cycle, as shown in Column 4 of Table 3. Thus, a combination of a low T'_g with the suboptimal fill depth required a drying cycle that may well have been uneconomical. Possibly, a reformulation might have led to a more efficient process.

12.7 *In situ* Water Release during Storage

A fully amorphous (glassy) freeze-dried product, consisting of 5% of a calcitonin gene-related protein (CGRP), 95% lactose and 3% residual water, was heated gently in a sealed DSC capsule. Its T_g is shown by the endothermic step in Scan 1 of Figure 4. Heating beyond T_g makes a material susceptible to more or less rapid physical and/or chemical

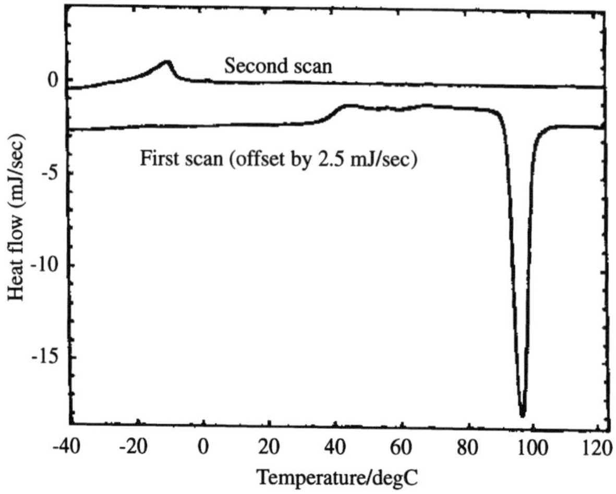

Figure 4 *DSC heating scans of formulated and freeze-dried CGRP (for details, see text). The first scan shows T_g, followed by an irruptive crystallisation of the stabiliser (lactose). The resulting product was recooled to −50°C and reheated. The scan shows a characteristic ice melting endotherm. Data from unpublished Pafra Biopreservation studies*

changes. In the preparation under test, such heating resulted in a sudden lactose crystallisation, leaving an amorphous product phase of 40% CGRP and 60% "residual" water. This is shown dramatically by the narrow, intense exotherm, representing the heat of lactose crystallisation in Scan 1 of Figure 4. The sample was cooled down to −50°C, annealed and rescanned (Scan 2). This scan shows no evidence of a glass transition but an ice melting endotherm, as typically observed for frozen solutions above their T_g. The sudden removal of the excipient by its crystallisation had thus left a residual (now dilute) CGRP/water solution with a subzero T_g, evidently below −50°C, rendering it unstable at ambient temperature. The broad endotherm is characteristic of ice melting into a concentrated solution.

12.8 Summary and Conclusions

The problems encountered and their resolution graphically illustrate the major impact made by changes of composition and fill depth on a previously optimised drying cycle, neglect or ignorance of which will almost invariably end in disaster. Since the practical freeze-drying activities and experiences of Pafra Biopreservation were largely confined to problem solving for individual client companies, we only saw "hospital cases", in the words of our Technical Manager. Nevertheless, from an archive that contains dozens of case studies, one obvious conclusion becomes evident: the process development staff of the client companies knew little or nothing of glass transitions or of the materials science approaches to the formulation of amorphous pharmaceutical solids. Even where the knowledge exists within a company, its significance does not generally reach the thinking of marketing departments or senior management. Too often the author has been invited by development departments to make presentations to senior managers, with the sole purpose of convincing them of modern thinking on freeze-drying. Unless this lack is remedied, freeze-drying will continue to be practised as a trial-and-error technology.

CHAPTER 13

Beyond Freeze-Drying

13.1 Alternative Water Removal Technologies

In contrast to freezing, followed by the sublimation of ice, water can be removed more directly by evaporation, as shown in Figure 1. Despite conventional wisdom, still widespread in the pharmaceutical industry, that freeze-drying is the preferred method of water removal, the alternatives deserve consideration. Under optimum conditions, the concentration step for suitably formulated labile products in vials containing approximately 1 ml solution, from an initial 5% to a final 95% solids, can be performed at room temperature in 2 h, and even less at elevated temperatures. For bulk volumes, spray drying offers an attractive alternative. The drying period of an individual droplet is then reduced to less than 10 s. Figure 2 shows, in diagrammatic fashion, the path of an isothermal drying process. Beyond the point of saturation solubility (S), the product is in a supersaturated state, and therefore *thermodynamically unstable*. On traversing the glass transition profile, the high viscosity will render the product *kinetically* stable. Stabilisation processes can be based on the optimisation of pressure/temperature/formulation conditions so as to minimise deterioration of a labile bioproduct while it traverses the supersaturated, unstable concentration range on its journey to the

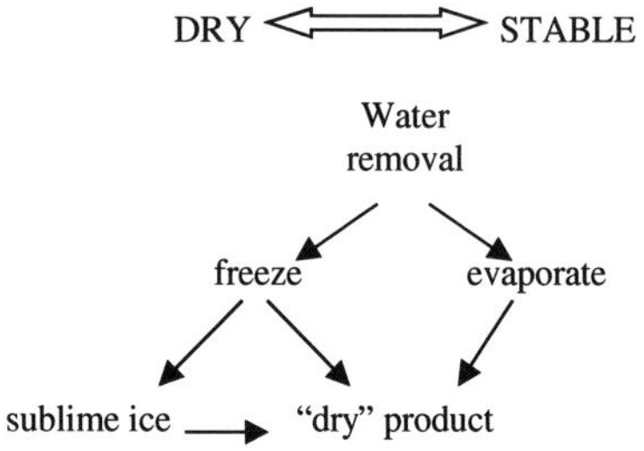

Figure 1 *Alternative pathways for the removal of water from aqueous solutions*

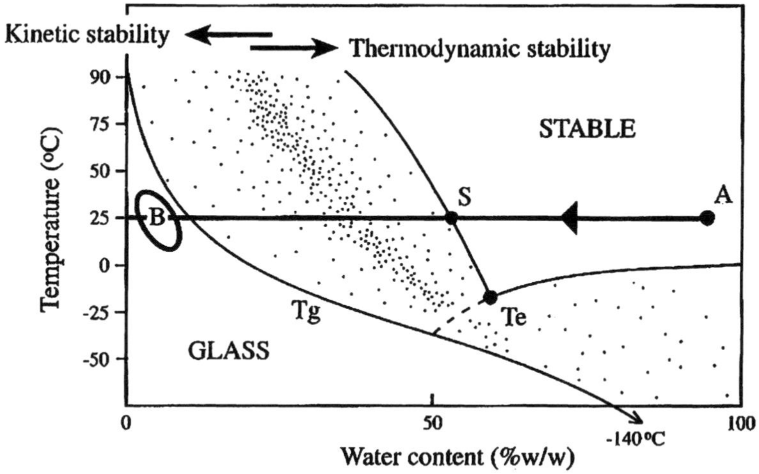

Figure 2 *Comparison of freeze-drying with evaporation of a dilute solution (A) that is formulated to yield an amorphous solid (B). The density of stippling is an indication of the probability of reversion to the equilibrium state (e.g. by crystallisation). For details, see text. Reproduced with permission from Franks*[162]

vitreous state. The Permazyme™ stabilisation process, developed in the author's company during the early 1990s, was based on the optimisation of pressure/temperature/formulation conditions, so as to achieve this aim.

13.2 Evaporative Drying: Principles and Practice

The drying process shown in Figure 3 represents the T_g/time relationship for a carbohydrate, containing 25% "residual" water, while it was being dried isothermally at 318 K under reduced pressure. The glass temperature corresponding to the drying temperature was reached after 90 min, beyond which drying of the glass still continued, but at a much reduced rate. Drying was judged to be complete after 6 h when the water content had decreased to 2%, with a corresponding product glass temperature of 360 K.

Depending on the processing method, the dried product can take the physical form of an opaque or transparent film, adhering to the containing vessel, or a fine powder. Just as for freeze-dried products, the stability under given storage conditions depends critically on the shape of the T_g/composition profile and on T_o, and the product is vulnerable to the same sudden, unpredictable and irreversible changes, if excursions

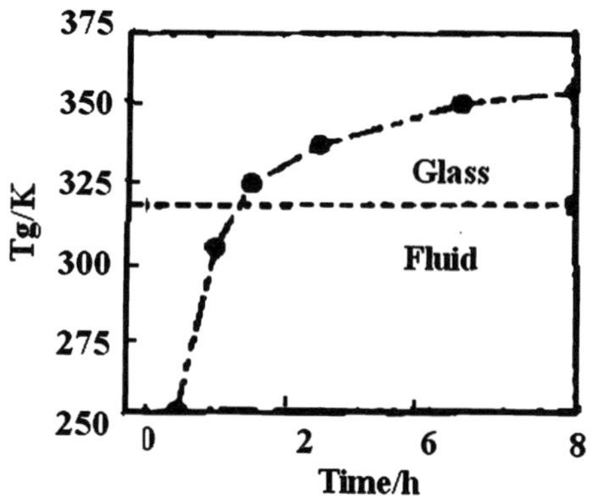

Figure 3 *T_g/time profile for a semi-dry PHC (polyhydroxy compound) product being dried in a vial under isothermal conditions (318 K)*

into the region of instability are permitted, either by a rise in the temperature or water sorption.

Contrary to conventional wisdom, the employment of ambient or elevated processing temperatures to remove water does *not* ncccssarily inactivate bioproducts, provided that care is taken in the design of appropriate formulations. The method has been successfully applied to the ambient temperature stabilisation of a wide range of formulated protein preparations.

13.3 Evaporative Drying: The Current State

Ambient temperature-stable DNA-modifying enzymes (restriction enzymes) that conventionally had to be shipped and stored in concentrated glycerol solution at –20 °C were produced in a shelf-stable format by the new evaporation technology and became available commercially.[160] An example in Figure 4 shows that such sensitive enzymes, dried and stored for extended periods, compare well with their freshly prepared counterparts. The technology was later applied to the dry-state stabilisation of microbial cultures for the dairy industry. Potentially, the most successful application of protein stabilisation has been the formulated insulin product Exubera, a spray-dried preparation, designed for delivery for inhalation; it was approved by the FDA. and the European Union in early 2006, some 15 years after the filing of the US patent,[161] for the treatment of Types 1 and 2 diabetes. Exubera combines the advantages of a shelf-stable product, even at tropical temperatures, with the replacement of delivery by injection.

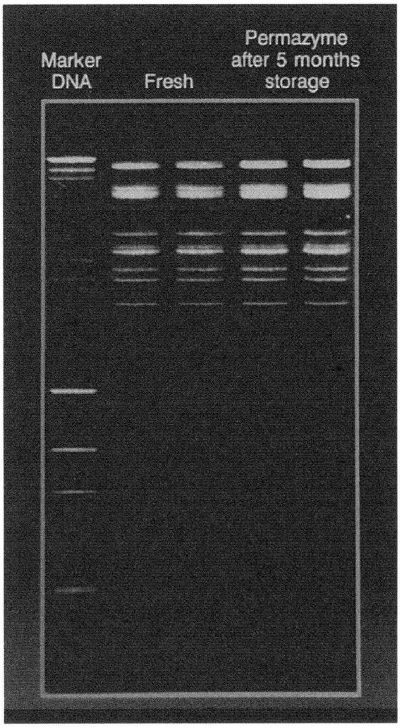

Figure 4 *Comparative gel electrophoresis patterns of fresh (in aqueous glycerol solution) PST 1 Restriction Enzyme produced digest with a "Ready-To-Go" dried preparation, after 5-month storage at ambient temperature (see Franks et al.)*[160]

13.4 Comparisons with Freeze-Drying

Advantages of evaporative drying over freeze-drying include:

(i) Capital equipment less expensive; freeze-drying equipment can be used.
(ii) Shorter processing time; lower energy requirements.
(iii) More reliable process monitoring and control (water removal does not rely on stochastic – uncontrollable – processes like ice nucleation).
(iv) Enhanced long-term stability at ambient, even elevated temperatures.
(v) Water is removed from the system permanently in a single step.
(vi) Spray drying lends itself to continuous processing.

Disadvantages of evaporative drying include:

(i) Necessity of powder filling may create sterility problems.
(ii) Spray drying requires large volumes of purified gas.

(iii) Drying in vials at ambient temperatures and reduced pressure is likely to yield transparent (invisible) product films, a distinct marketing disadvantage.

To the author's knowledge, the evaporative drying alternative is still waiting to be exploited on a commercial scale by the pharmaceutical industry. The reasons for this conservatism are several and may be mainly related to questions of safety and regulatory requirements. This is understandable but it should not deter processors of solid-state pharmaceuticals from a study of the technology. There can be little doubt that large scale freeze-drying, as currently practised, must eventually give way to more efficient and economic means of water removal.

Acknowledgements

This book has been many, too many, years in gestation. Colleagues and friends who have helped, consciously or subconsciously, directly or indirectly, with its eventual birth are too numerous for me to include them all by name. I may even have forgotten some of them, for which I apologise in advance. Put it down to intermittent memory failure. Acknowledgements might be listed in alphabetical order or by the significance of individual contributions (consciously or unconsciously). They might be listed in order of seniority, because they range from grey-haired, thoughtful professors to long-haired, brash undergraduate students. I have however decided on chronology, which begins in 1966, at the Unilever Research Laboratory near Bedford, where I first encountered the words "stability" and "shelf life". As someone who had never attended a single life science course, I am indebted to my then boss Tony James, FRS, and colleagues, such as Mike Tombs, and Lindsay Morris who patiently explained the mysteries of biology and biochemistry, in particular the elements of protein, lipid and polysaccharide biochemistry. I well remember the coffee breaks in the lab, when they all patiently fielded questions from a physical scientist, who could hardly distinguish a mitochondrion from a chloroplast. Geoff Telling had the analytical skills of chasing and detecting nasty residues in foods, at parts of per million or billion levels, and he has remained a friend to this day. I quickly had to learn "all" about plants and microbiology, for which I thank Laurie Jones and Grahame Gould; both of them spent much time with the beginner.

I first entered the realm of "preservation" in 1974, when it became my lot to manage a major project, the aim being the *in vitro* propagation of oil palm tissue. It was my first excursion into anything connected with live cells and tissues. Freezing seemed the obvious way forward, and so I entered the then highly politicised realm of cryopreservation. To this end, I joined the Society of Cryobiology and in due course even made it to the Board of Governors. I am still indebted to several of its members for teaching me all about low-temperature methods of preservation. I am particularly grateful to several of the Society's founders: Harry Meryman, the late Peter Steponkus, Peter Mazur, John Baust and David Pegg.

It did not take me too long to realise that freezing was not the method for the preservation of oil palm tissue and also that ice is an efficient killer. At the same time, we knew that low temperature did retard deleterious reactions, perhaps even those associated with senescence. Like many physicists, when the "real world" became too complicated, we escaped to "model systems", such as isolated proteins, to study the effects of freezing. It was then that I had the opportunity of spending a short sabbatical period in the University of Cambridge's Department of Plant Sciences, or Botany, as it was then called. Here I made the acquaintance of Patrick Echlin and Helen Skaer, who taught me the basics of optical and electron microscopy. Together we explored the phenomena of freezing and vitrification of water. On the way, I also took lessons in plant biochemistry from the late Tom ap Rees, who became Department Head. He was the best University lecturer I have ever encountered. The short collaboration with Patrick and Helen first convinced me that undercooled and vitreous water might be the way to successful preservation. Three years later, when I said good bye to Unilever, or *vice versa*, I rejoined my former Cambridge partners on a more permanent basis, and so began my 22-year stay in Cambridge and a gradual drift towards physiological and pharmaceutical sciences. I established a group devoted to the study of *in vivo* preservation against freezing. After several false starts, it was my good fortune to appoint Sheila Mathias as my main helper and general factotum. I thank her for teaching me the basics of cell and tissue culture and for keeping the lab on an even keel, no mean achievement in the Cambridge environment. Together we established our group as international experts in novel biopreservation techniques. We were helped by the proximity of the Medical Research Council's Cryobiology Group, not only by helpful discussions with David Pegg and Mike Taylor, but also by regular supplies of liquid nitrogen. A stroke of luck brought us the visit of Harry Levine from General Foods in New York, and so began a deep friendship and a most fruitful scientific collaboration. It was he and his partner, Louise Slade, who immediately appreciated the significance of the amorphous state and glasses, as applied to food processing. "Harry and Louise" became a byword for all forward-looking food technologists. I admit that I did not have the same level of instant success and enlightenment with the pharmaceutical process industry.

Our researches into the mechanisms of biostability eventually led to the establishment of Pafra Biopreservation as a start-up venture on the

Cambridge Science Park. We received help and encouragement from John Bradfield, the Senior Bursar of Trinity College, who had founded the Science Park. I owe a special debt of gratitude to the late Max Kochmann, Chairman of Pafra Ltd., for his faith in the potential of our technology and for seed finance. It enabled us to create a business, devoted to the achievement and marketing of "Stability", by applying *in vitro* the mechanisms that plants, microorganisms, insects and other species developed long ago to survive the different forms of water stress. Eventually Sheila decided to give up working, and I was fortunate to find a valuable replacement in Tony Auffret as Technical Manager; he brought with him a fount of experience, and we have remained friends to this day. Begun in 1990 to attract revenue, we developed a programme of freeze-drying courses, which are still being presented on a regular basis in several European countries. Some of the subject matter discussed in this book has been taken from our course manual, after suitable modification and updating. Tony should therefore be regarded almost as a co-author, and he deserves my special thanks.

Pafra Biopreservation was a strange hybrid. While the salaried staff was engaged in developing a freeze-drying consultancy business, in order to pay our salaries, we relied for progressing our own stabilisation technology on a series of visitors who had read about our achievements and were keen to join us in our research. They formed a valuable academic adjunct group, without which we could not have succeeded. They came from Japan, USA, The Netherlands, Russia, Cuba and Cambridge. Their proximity and experience served as inspiration to our own staff, and they helped out in times of crisis. Their names are to be found scattered throughout the bibliography of this book.

Others to whom I owe thanks, both for professional help and advice, and also for firm friendship, include Rainer Jaenicke (then Professor of Biophysics at the University of Regenburg), Stephan Fischer (Boehringer-Mannheim, now Sandoz, Penzberg, Germany), Chris Körber (European Patent Office, Munich, Germany), Mike Pikal, the "king of freeze-drying", the late Donald Irish (Waterloo University, Canada), Steve Hurst, Linda Foster and David Lechuga (all at Inhale Therapeutic Systems, now renamed Nektar Therapeutics, San Carlos, CA). And there are or have been others, too numerous to mention them all, who helped our Stabilisation enterprise along. They all became members of what my wife still refers to as "your international scientific mafia". According to the proverb "A rolling stone gathers no moss", but in my personal experience, the opposite has been true: I have done much rolling but gathered a mass of valuable moss.

Bibliography

References

1. S. Purchas, 1596. The first navigation of William Barents, alias Bernards, into the North Seas. Written by Great de Veer. In *Hakluytus Posthumus or Purchas his Pilgrimes*, **13**, 35. MacLehose & Sons, Glasgow, 1906.
2. J.de. Acosta, 1590. *Historia natural y moral de las Indias* (translat. E. Grimston), London, 1604. Haklyuyt Society reprint, 1880. Scvillc, Spain.
3. E. Chain, S.A. Goldblith, L. Rey and A. Spicer, Food technology in the 1980s, *Proc. Roy. Soc. B*, 1975, **191**, 1–198.
4. E.W. Flosdorf, *Freeze-Drying*, Reinhold Publishing Corp., New York, 1949.
5. G.-W. Oetjens, *Gefriertorcknen*, VCH, Weinheim, Germany, 1997.
6. T.A. Jennings, *Lyophilization – Introduction and Basic Principles*, Interpharm Press, Englewood, CO, 1999.
7. P. Cameron, (ed.), *Good Pharmaceutical Freeze-Drying Practice*, Interpharm Press, Buffalo Grove, IL, 1997.
8. L. Rey and J.C. May, (eds), *Freeze-Drying/Lyophilization of Pharmaceutical and Biological Products*, Marcel Dekker, Inc., New York, 1999.
9. H. Levine and L. Slade, Water as plasticiser: physico-chemical aspects of low-moisture polymeric systems, *Water Sci. Rev.*, 1988, **3**, 79–185.
10. V.V. Mozhaev and E. Martinek, Inactivation and reactivation of proteins, *Mol. Biol. (Moscow)*, 1982, **16**, 676–694.
11. K. Eichner, The influence of water content on non-enzymic browning reactions in dehydrated foods and model systems and the inhibition of fat oxidation by browning intermediates, In *Water Relations in Foods*, R.B. Duckworth, (ed.), Academic Press, London, 1975, pp. 417–434.
12. J. Büchner and R. Rudolph, Renaturation, purification and characterization of recombinant AB fragments produced in *E. coli*. *Biotechnology* 1991, **9**, 97–162.
13. H. Willemer, In *Freeze-Drying/Lyophilization of Pharmaceutical and Biological Products*, L. Rey and J. May (eds), Marcel Dekker, Inc., New York, 1999, p. 105ff.

14. R.H.M. Hatley, F. Franks and S.F. Mathias, The stabilisation of labile biochemicals by undercooling, *Process. Biochem.*, 1987, **22**(12), 169–172.
15. P. Douzou, C. Balny and F. Franks, New trends in cryoenzymology. I. Supercooled aqueous solutions, *Biochimie*, 1978, **60**, 151–158.
16. F. Franks, S.F. Mathias, P. Galfre, S.D. Webster and D. Brown, Ice nucleation and freezing in undercooled cells, *Cryobiology*, 1983, **20**, 298–309.
17. F. Franks, *Biophysics and Biochemistry at Low Temperatures*, Cambridge University Press, 1985.
18. F. Franks, Nucleation of ice and its management in ecosystems, *Philos. Trans. Roy. Soc. London. A*, 2003, **361**, 557–574.
19. D. Turnbull and J.C. Fisher, Rate of nucleation in condensed systems, *J. Chem. Phys.*, 1949, **17**, 71–73.
20. P.V. Hobbs, *Ice Physics*, Oxford University Press, 1974.
21. C.A. Angell, In: *Water – A Comprehensive Treatise*, F. Franks (ed.), Chapter 1, 1982, pp. 1–81.
22. S.F. Mathias, F. Franks and K. Trafford, Nucleation and growth of ice in deeply undercooled erythrocytes, *Cryobiology*, 1984, **21**, 123–132.
23. R.W. Michelmore and F. Franks, Nucleation rates of ice in undercooled water and aqueous solutions of polyethylene glycol, *Cryobiology*, 1982, **19**, 163–171.
24. F. Franks, S.F. Mathias and K. Trafford, The nucleation of ice in undercooled water and aqueous polymer solutions, *Colloid. Surf.*, 1984, **11**, 275–285.
25. J.O. Krog, K.E. Zachariassen, B. Larsen and O. Smidsrød, Buffering in Afro-Alpine plants due to nucleating agent-induced water freezing, *Nature*, 1979, **282**, 300–301.
26. N. Murase, M. Ruike, N. Matsunaga, M. Hayakawa, Y. Kaneko and Y. Ono, Spider silk has an ice nucleating activity, *Naturwissenschaften*, 2001, **88**, 117–118.
27. P.W. Wilson, A.F. Heneghan and A.D.J. Haymet, Ice nucleation in nature: supercooling point (SCP) measurements and the role of heterogeneous nucleation, *Cryobiology*, 2003, **46**, 88–98.
28. C. Körber, Phenomena at the advancing ice–liquid interface; solutes, particles and biological cells, *Quart. Rev. Biophys.*, 1988, **21**, 229–298.
29. N.H. Fletcher, *The Chemical Physics of Ice*, Cambridge University Press, 1970.
30. G.P. Johari, A. Hallbrücker and E. Mayer, The glass–liquid transition of hyperquenched water, *Nature*, 1987, **330**, 552–553.
31. P. Echlin, (ed.), *Low-Temperature Microscopy and Analysis*, Plenum Press, New York, 1992.
32. F. Franks, M.H. Asquith, H.LeB. Skaer and B. Roberts, Aggregation patterns and microstructure in aqueous polymer solutions: comparison of quench-fracture electron microscopy with predictions based on rheological measurements, *CryoLetters*, 1979, **1**, 104–113.

33. J.L. Finney and A.K. Soper, Solvent structure and perturbation in solutions of chemical and biological importance, *Chem. Soc. Rev.*, 1994, **23**, 1–10.
34. F. Sussich, R. Urbani, F. Princivalle and A. Cesaro, Polymorphic amorphous and crystalline forms of trehalose, *J. Am. Chem. Soc.*, 1998, **120**, 7893–7899.
35. K. Takaizumi, Liquid–solid phase diagram of PrOH/water and BuOH/water systems as studied by DSC, *J. Solution Chem.*, 2000, **29**, 377–388.
36. K. Takaizumi and T. Wakabayashi, The freezing process in methanol–, ethanol–, and propanol–water systems as revealed by DSC, *J. Solution Chem.*, 1997, **26**, 927–939.
37. O. Mishima and H.E. Stanley, The relationship between liquid, supercooled and glassy water, *Nature*, 1998, **396**, 4199–4202.
38. O. Mishima and Y. Suzuki, Vitrification of emulsified liquid water under pressure, *J. Chem. Phys.*, 2001, **115**, 4199–4202.
39. F. Franks, Scientific and technological aspects of aqueous glasses, *Biophys. Chem.*, 2003, **105**, 251–261.
40. M.J. Pikal and D.R. Rigsbee, The stability of insulin in crystalline and amorphous solids: observations of greater stability for the amorphous form, *Pharm. Res.*, 1997, **14**, 1379–1387.
41. F. Franks and R.H.M. Hatley, Low temperature unfolding of chymotrypsinogen, *CryoLetters*, 1985, **6**, 171–180.
42. A.P. MacKenzie, Non-equilibrium freezing behaviour of aqueous systems, *Philos. Trans. Roy. Soc. B*, 1977, **278**, 167–188.
43. F. Franks, In *Water – A Comprehensive Treatise*, F. Franks (ed.), Vol. 7, Plenum Press, New York, 1982, pp. 216–338.
44. F. Franks, *Water – A Matrix of Life*, Royal Society of Chemistry, Cambridge, 2000.
45. A. Calvelo, *Recent studies on meat freezing*, In *Developments in Meat Science*, R. Lawrie (ed.), Vol. 2, Applied Science Publishers, London, 1981, pp. 125–158.
46. W. Burchard, *Solution thermodynamics of non-ionic water soluble polymers*, In *Chemistry and Technology of Water Soluble Polymers*, C.A. Finch (ed.), Plenum Press, New York, 1983, pp. 125–142.
47. K. Izutsu and S. Kojima, Freeze-concentration separates proteins and polymer excipients into different amorphous phases, *Pharm. Res.*, 2000, **17**, 1316–1322.
48. J.F. Brandts, The thermodynamics of protein denaturation, *J. Amer. Chem. Soc.*, 1964, **86**, 4291–4301.
49. F. Franks, Protein destabilization at low temperatures, *Adv. Protein Chem.*, 1995, **46**, 105–139.
50. S. Bociek and F. Franks, Proton exchange in aqueous solutions of glucose – hydration of carbohydrates, *J. Chem. Soc. Faraday Trans. I*, 1979, **75**, 262–270.

51. F. Franks, R.H.M. Hatley and H.L. Friedman, The thermodynamics of protein stability. Cold denaturation as a general phenomenon, *Biophys. Chem.*, 1988, **31**, 307–315.
52. D.G. Archer, Thermodynamic properties of the NaCl + H_2O system. II. Thermodynamic properties of NaCl(aq), NaCl.$2H_2O$(cr), and phase equilibria, *J. Phys. Chem. Ref. Data*, 1992, **21**, 793–829.
53. F.S. Soliman and L. Van den Berg, Factors affecting freezing damage of lactate dehydrogenase, *Cryobiology*, 1971, **8**, 73–78.
54. J.E. Lovelock, The haemolysis of human red blood cells by freezing and thawing, *Biochim. Biophys. Acta*, 1953, **10**, 414–426.
55. L. Van den Berg and D. Rose, The effects of freezing on pH and composition of sodium and potassium phosphate solutions. The reciprocal system KH_2PO_4–Na_2HPO_4–H_2O, *Arch. Biochem. Biophys.*, 1959, **81**, 319–329.
56. N. Murase and F. Franks, Salt precipitation during the freeze-concentration of phosphate buffer solutions, *Biophys. Chem.*, 1989, **34**, 293–300.
57. L. Yu, N. Milton, E.G. Groleau, D.S. Misra and R.E. Vansickle, Existence of a mannitol hydrate during freeze drying and practical implications, *J. Pharm. Sci.*, 1999, **88**, 196–198.
58. T.E. Kiovsky and R.E. Pincock, Mutarotation of glucose in frozen aqueous solutions, *J. Am. Chem. Soc.*, 1966, **88**, 7704–7710.
59. R.H.M. Hatley, F. Franks and H. Day, Subzero temperature preservation of reactive fluids in the undercooled state. 2. The effect on the oxidation of ascorbic acid of freeze concentration and undercooling, *Biophys. Chem.*, 1986, **24**, 187–192.
60. O. Fennema, *Activity of enzymes in partially frozen aqueous systems*, In *Water Relations of Foods*, R.B. Duckworth (ed.), Academic Press, London, 1975, pp. 397–413.
61. J.F. Brandts, J. Fu and J.H. Nordin, The low temperature denaturation of chymotrypsinogen in aqueous solution and in frozen aqueous solution, In *The Frozen Cell*, G.E.W. Wolstenholme and M. O'Connor (eds.), J.A. Churchill, London, 1970, pp. 189–208.
62. D.R. MacFarlane, Physical aspects of vitrification in aqueous solutions, *Cryobiology*, 1987, **24**, 181–195.
63. F. Franks and C. van den Berg, *Topics in Pharmaceutical Sciences*, D.J.A. Crommelin and K.K. Midha (eds.), Vol. 7, Medpharm Scientific Publishers, Stuttgart, 1991, p. 215.
64. F. Franks, Solid aqueous solutions, *Pure Appl. Chem.*, 1993, **65**(12), 2527–2537.
65. V. Zepharovich, Die Kristallformen des Mannit, *Z. F. Kryst.*, 1888, **13**, 145–149.
66. A. Burger, J.-O. Henck, S. Hetz, J.M. Rollinger, A.A. Weissnicht and H. Stöttner, Energy/temperature diagram and compression behaviour of the polymorphs of D-mannitol, *J. Pharm. Sci.*, 2000, **89**, 457–468.

67. C. Telang, R. Suryanarayanan and L. Yu, Crystallization of D-mannitol in binary mixtures with NaCl: phase diagram and polymorphism, *Pharm. Res.*, 2003, **20**, 1939–1945.
68. M.J. Akers, N. Milton, S.R. Byrn and S.L. Nail, Glycine crystallization during freezing: the effect of salt form, pH and ionic strength, *Pharm. Res.*, 1995, **12**, 1457–1461.
69. G.A. Jeffrey, Conformational studies in the solid state: extrapolations to molecules in solution, *Adv. Chem. Ser.*, 1973, **117**, 177–196.
70. G.A. Jeffrey and D.-B. Huang, The hydrogen bonding in the crystal structure of raffinose pentahydrate, *Carbohyd. Res.*, 1990, **206**, 173–182.
71. F. Franks, Freeze drying: from empiricism to predictability, *CryoLetters*, 1990, **11**, 93–110.
72. D.P. Miller, J.J. de Pablo and H. Corti, Thermophysical properties of trehalose and its concentrated aqueous solutions, *Pharm. Res.*, 1997, **14**, 578–590.
73. H. Nicolajsen and A. Hvidt, Phase behaviour of the system trehalose–NaCl–water, *Cryobiology*, 1994, **31**, 199–205.
74. J.L. Green and C.A. Angell, Phase relations and vitrification in saccharide–water solutions and the trehalose anomaly, *J. Phys. Chem.*, 1989, **93**, 2880–2882.
75. M. Yoshioka, B.C. Hancock and G. Zografi, Crystallization of indomethacin from the amorphous state below and above its glass transition temperature, *J. Pharm. Sci.*, 1994b, **83**, 1700–1705.
76. C.A. Angell, The glass transition, *Curr. Opin. Solid State Mater. Sci.*, 1996, **1**, 578–585.
77. R.S. Smith and B.D. Kay, The existence of supercooled liquid water at 150 K, *Nature*, 1999, **398**, 788–791.
78. L. Finegold, F. Franks and R.H.M. Hatley, Glass/rubber transitions and heat capacities of binary sugar blends, *J. Chem. Soc. Faraday Trans. I*, 1989, **85**, 2945–2951.
79. R.H. Tromp, R. Parker and S.G. Ring, A neutron scattering study of the structure of amorphous glucose, *J. Chem. Phys.*, 1997, **107**, 6038–6049.
80. E. Caffarena and R.J. Grigera, Crystal, melted and glassy states of glucose. A molecular dynamics simulation, *J. Chem. Soc. Faraday Trans.*, 1996, **92**, 2285–2289.
81. P.B. Conrad and J.J. de Pablo, Computer simulation of the cryoprotectant disaccharide α,α-trehalose in aqueous solution, *J. Phys. Chem. A*, 1999, **103**, 4049–4055.
82. C.J. Roberts and P.G. Debenedetti, Structure and dynamics in concentrated amorphous carbohydrate–water systems by molecular dynamics simulation, *J. Phys. Chem. B*, 1999, **103**, 7308–7318.
83. B.C. Hancock, E.Y. Shalaev and S.L. Shamblin, Polymorphism: a pharmaceutical science perspective, *J. Pharm. Pharmacol.*, 2002, **54**, 1151–1152.

84. J.L. Finney, What's so special about water?, *Philos. Trans. Roy. Soc. B.*, 2004, **359**, 1145–1165.
85. L. Pauling, The structure and entropy of ice and other crystals with some randomnent of atomic arrangementsm, *J. Am. Chem. Soc.*, 1935, **57**, 2680–2684.
86. P.G. Debenedetti and F.H. Stillinger, Supercooled liquids and the glass transition, *Nature*, 2001, **410**, 259–267.
87. P.G. Debenedetti, *Metastable Liquids*, Princeton University Press, 1996.
88. C.T. Moynihan, Structural relaxation and the glass transition, *Rev. Mineral.*, 1995, **32**, 1–19.
89. S.P. Apte and S.O. Ogwu, A review and classification of emerging excipients in parenteral medications, *Pharm. Technol. Europe*, 2003, **15**(12), 45–52.
90. P. Crowley and L. Martini, Drug–excipient interactions, *Pharm. Technol. Europe*, 2001, **13**, 26–34.
91. R.C. Moreton, Aspects relating to excipient quality and specifications, *Pharm. Technol. Europe*, 1999, **11**(12), 26–31.
92. J. Michaud and G. Haest, Erythritol: a new multipurpose excipient, *Pharm. Technol. Europe*, 2003, **15**(10), 69–72.
93. T. Auffret, *Rational formulation design – can the regulators be educated?* In *Amorphous Food and Pharmaceutical Systems*, H. Levine (ed.), Royal Society of Chemistry, Cambridge, 2002, pp. 187–199.
94. K. Komai and N. Murase, Water sorption behaviour of glycinebetaine and the state diagram of its aqueous system, In *Water Properties of Food, Pharmaceutical, and Biological Materials*, M. del Pilar Buera, et al., (eds), Taylor & Francis, Boca Raton, FL, 2006, pp. 647–654.
95. M. Gordon and J.S. Taylor, Ideal copolymers and the second order transitions of synthetic rubber. 1. Non-crystalline copolymers, *J. Appl. Chem.*, 1952, **2**, 493–500.
96. M. Sugisaki and H. Suga, Calorimetric study of glassy state and heat capacities of glassy water and cubic ice, *Bull. Chem. Soc. Japan*, 1968, **41**, 2591–2599.
97. V. Velikov, S. Borick and C.A. Angell, The glass transition of water, based on hyperquenching experiments, *Science*, 2001, **294**, 2335–2338.
98. V. Truong, B.B. Bhandari, T. Howes and B. Adhikari, Analytical model for the prediction of glass transition temperature of food systems, In *Amorphous Food and Pharmaceutical Systems*, H. Levine (ed.), Royal Society of Chemistry, Cambridge, 2002, pp. 31–47.
99. F. Franks, Conformational stability of proteins, In *Protein Biotechnology*, F. Franks (ed.), Humana Press, Totowa, NJ, 1993, pp. 395–436.
100. B.J. Aldous, A.D. Auffret and F. Franks, The crystallization of hydrates from amorphous carbohydrates, *CryoLetters*, 1995, **16**, 181–186.
101. J.F. Carpenter and J.H. Crowe, Infrared spectroscopic studies on the interaction of carbohydrates with dried proteins, *Biochemistry*, 1989, **28**, 3916–3922.

102. A.W. Ford and Z. Allahiary, The adverse effect of glycation of human serum albumin on its preservative activity in the freeze-drying and accelerated degradation of alkaline phosphatase, *J. Pharm. Pharmacol.*, 1993, **45**, 900–906.
103. D.M. Bunk, Characterization of the glycation of albumin in freeze-dried and frozen human serum, *Anal. Chem.*, 1997, **69**, 2457–2463.
104. L. Slade and H. Levine, Beyond water activity: recent advances based on an alternative approach to the assessment of food quality and safety, *Crit. Rev. Food Sci. Nutr.*, 1991, **30**, 115–360.
105. E.Yu. Shalaev, F. Franks and P. Echlin, Crystalline and amorphous phases in the ternary system water–sucrose–sodium chloride, *J. Phys. Chem.*, 1996, **100**, 11.
106. Q. Lu and G. Zografi, Properties of citric acid at the glass transition, *J. Pharm. Sci.*, 1997, **86**, 1374–1378.
107. G. Steinbach, *Equations for the Heat and Mass Transfer in Freeze-Drying of Porous and Non-porous Layers and Bodies*, International Institute of Refrigeration (XIII), Washington, 1971, pp. 674–683.
108. M.J. Pikal, M.L. Roy and S. Shah, Mass and heat transfer in vial freeze-drying of pharmaceuticals – role of the vial, *J. Pharm. Sci.*, 1984, **73**, 1224–1237.
109. D.E. Overcashier, T.W. Patapoff and C.C. Hsu, Lyophilization of protein formulations in vials. Investigation of the relationship between resistance to vapor flow during primary drying and small-scale product collapse, *J. Pharm. Sci.*, 1999, **88**, 688–695.
110. W. Umrath, Berechnung von Gefriertrocknungszeiten für die elektronenmikroskopische Präparation, *Mikroskopie Wien*, 1983, **40**, 9–37.
111. B.S. Chang and N.L. Fisher, Development of an efficient single-step freeze-drying cycle for protein formulations, *Pharm. Res.*, 1995, **12**, 831–837.
112. K. Kasraian and P.P. DeLuca, The effect of tertiary butyl alcohol on the resistance of the dry product layer during primary drying, *Pharm. Res.*, 1995, **12**, 491–495.
113. J. Oesterle, F. Franks and T. Auffret, The influence of tertiary butyl alcohol and volatile salts on the sublimation of ice from frozen solutions: implications for freeze-drying, *Pharm. Dev. Technol.*, 1998, **3**, 175–183.
114. E.Yu. Shalaev and F. Franks, State of model mixtures during freezing and drying. Impact on product quality. *Cryobiology* 1996, 33, 14–26.
115. K. Murgatroyd, In *Good Pharmaceutical Freeze-Drying Practice*, P. Cameron (ed.), Interpharm Press, Buffalo Grove, IL, 1997, pp. 1–154.
116. R.H.M. Hatley, C. van den Berg and F. Franks, The unfrozen water content of maximally freeze concentrated carbohydrate solutions – validity of the methods used for its determination, *CryoLetters*, 1991, **12**, 113–126.
117. R.H.M. Hatley and A. Mant, Determination of the unfrozen water content of maximally freeze-concentrated carbohydrate solutions, *Int. J. Biol. Macromol.*, 1993, **15**, 227–232.

118. M.J. Pikal, S. Shah, M.L. Roy and R. Putman, The secondary drying stage of freeze-drying – drying kinetics as a function of temperature and chamber pressure, *Int. J. Pharm.*, 1990, **60**, 203–217.
119. B.J. Aldous, F. Franks and A.L. Greer, Diffusion of water within an amorphous carbohydrate, *J. Mater. Sci.*, 1997, **32**, 301–308.
120. E.Yu. Shalaev and F. Franks, Structural glass transitions and thermophysical processes in amorphous carbohydrates and their supersaturated solutions, *J. Chem. Soc. Faraday Trans.*, 1995, **91**, 1511–1517.
121. E.Yu. Shalaev, 1991. Ph.D. Thesis, Institute of Molecular Biology, Koltsovo, Russia.
122. M.J. Millman, A.I. Liapis and J.M. Marchello, Note on the economics of batch freeze dryers, *J. Food Technol.*, 1985, **20**, 541–551.
123. L.D. Butler, Regulatory issues: an American perspective, In *Good Pharmaceutical Freeze-Drying Practice*, P. Cameron (ed.), Interpharm Press, Inc., Buffalo Grove, IL, USA, 1997, pp. 171–186.
124. P. Monger, Regulatory issues: a European perspective, In *Good Pharmaceutical Freeze-Drying Practice*, P. Cameron (ed.), Interpharm Press, Inc., Buffalo Grove, IL, USA, 1997, pp. 155–170.
125. E.Yu. Shalaev and F. Franks, Solid–liquid state diagrams in pharmaceutical lyophilisation: crystallisation of solutes, In *Amorphous Food and Pharmaceutical Systems*, H. Levine (ed.), Royal Society of Chemistry, Cambridge, 2002, pp. 200–215.
126. T. Suzuki and F. Franks, Solid–liquid phase transitions and amorphous states in ternary sucrose–glycine–water systems, *J. Chem. Soc. Faraday Trans.*, 1993, **89**, 3283–3288.
127. L. Slade and H. Levine, Structural stability of intermediate moisture foods – a new understanding?, In *Food Structure – Its Creation and Evaluation*, J.M.V. Blanshard and J.R. Mitchell (eds.), Butterworths, London, 1988, pp. 115–147.
128. H. Levine, (ed.), *Amorphous Food and Pharmaceutical Systems*, Royal Society of Chemistry, Cambridge, 2002.
129. G. Buckton and P. Darcy, The influence of additives on the recrystallisation of amorphous spray dried lactose, *Int. J. Pharm.*, 1995, **121**, 81–87.
130. R. Haikala, R. Eerola, V.P. Tanninen and J. Yliruusi, Polymorphic changes of mannitol during freeze-drying: effect of surface-active agents, *PDA J. Pharm. Sci. Technol.*, 1997, **51**, 96–101.
131. G. Buckton, E. Yonemochi, J. Hammond and A. Moffat, The use of near-infrared spectroscopy to detect changes in the form of amorphous and crystalline lactose, *Int. J. Pharm.*, 1998, **168**, 231–241.
132. F.P.A. Fabbiani, D.R. Allan, A. Dawson, W.I.F. David, P.A. McGregor, I.D.H. Oswald, S. Parsons and C.R. Pulham, Pressure-induced formation of a solvate of paracetamol, *Chem. Commun.*, 2003, 3004–3005.
133. W. Kauzmann, The nature of the glassy state and the behavior of liquids at low temperatures, *Chem. Rev.*, 1948, **43**, 219–256.

134. H. Vogel, The law of the relation between the viscosity of liquids and the temperature, *Physik. Z.*, 1921, **22**, 645–646.
135. G.S. Fulcher, Analysis of recent measurements of the viscosity data for concentrated polymer solutions, *J. Am. Ceram. Soc.* 1925, 8, 339–355, 789–794.
136. D. Turnbull and M.H. Cohen, Free-volume model of the amorphous phase: glass transition, *J. Chem. Phys.*, 1961, **34**, 120–125.
137. C.A. Angell, Structural instability and relaxation in liquid and glassy phases near the fragile liquid limit, *J. Non-Cryst. Solids*, 1988, **102**, 205–221.
138. G. Adams and J.H. Gibbs, Temperature dependence of cooperative relaxation properties in glass-forming liquids, *J. Chem. Phys.*, 1965, **43**, 139–146.
139. R.L. Remmele and W.R. Gombotz, Differential scanning calorimetry: a practical tool for elucidating the stability of liquid biopharmaceuticals, *Biopharm Europe*, 2000, **12**, 56–65.
140. A. Boller, C. Schick and B. Wunderlich, Modulated differential scanning calorimetry in the glass transition region, *Thermochim. Acta*, 1995, **266**, 97–111.
141. C.T. Moynihan, A.J. Eastel, J. Wilder and J. Tucker, Dependence of glass transition temperature on heating and cooling rate, *J. Phys. Chem.*, 1974, **78**, 2673–2677.
142. B.C. Hancock, C.R. Dalton, M.J. Pikal and S.L. Shamblin, A pragmatic test of a simple calorimetric method for determining the fragility of some amorphous pharmaceutical materials, *Pharm. Res.*, 1998, **15**, 762–767.
143. C.T. Moynihan, S.-K. Lee, M. Tatsumisago and T. Minami, Estimation of activation energies for structural relaxation and viscous flow from DTA and DSC experiments, *Thermochim. Acta*, 1996, **280/281**, 153–162.
144. M.L. Williams, R.F. Landel and J.D. Ferry, The temperature dependence of relaxation mechanisms in amorphous polymers and other glass-forming liquids, *J. Am. Chem. Soc.*, 1955, **77**, 3701–3707.
145. J.L. Ford and P. Timmins, *Pharmaceutical Thermal Analysis: Techniques and Applications*, Ellis Horwood Ltd, Chichester, UK, 1989.
146. C. Ahlneck and G. Zografi, The molecular basis of moisture effects on the physical and chemical stability of drugs in the solid state, *Int. J. Pharm.*, 1990, **62**, 87–95.
147. H. Ahmed, G. Buckton and D.A. Rawlins, The use of isothermal microcalorimetry in the study of small degrees of amorphous content of a hydrophobic powder, *Int. J. Pharm.*, 1996, **130**, 195–201.
148. R.G. Strickley and B.D. Anderson, Solid state stability of human insulin, *Pharm. Res.*, 1996, **13**, 1142–1153.
149. D. Girlich, 1991. Multikernresonanzuntersuchungen zur molekularen Dynamik wässeriger Saccharidlösungen. [Study of the molecular dynamics in aqueous saccharide solutions by multi-nuclear resonance methods.]

Ph.D. Thesis, University of Regensburg; translated in summary form into English by F. Franks.

150. D. Girlich and H.-D. Lűdemann, Molecular mobility of sucrose in aqueous solution studied by ^{13}C NMR relaxation, *Z. Naturforsch.*, 1993, **48c**, 407–413.
151. D. Girlich, H.-D. Lűdemann, C. Buttersack and K. Buchholz, c,T-dependence of the self diffusion in concentrated aqueous sucrose solutions, *Z. Naturforsch.*, 1994, **49c**, 258–264.
152. S. Ablett, A.H. Clark, M.J. Izzard and P.J. Lillford, Modelling of heat capacity–temperature data for sucrose–water systems, *J. Chem. Soc. Faraday Trans.*, 1992, **88**, 795–802.
153. J.C. May, R.M. Wheeler, N. Etz and A. Del Grosso, Measurement of final container residual moisture in freeze-dried biological products, *Dev. Biol. Stand.*, 1992, **74**, 153–164.
154. J.C. May, Regulatory control of freeze-dried products: Iimportance and evaluation of residual moisture, In *Freeze-Drying/Lyophilization of pPharmaceutical and bBiological pProducts*, L. Rey and J.C. May (eds.), Marcel Dekker Inc, New York, 1999.
155. L. Streefland, A.D. Auffret and F. Franks, Bond cleavage reactions in solid aqueous carbohydrate solutions, *Pharm. Res.*, 1998, **15**, 843–849.
156. D. Girlich and H.-D. Lűdemann, Molecular mobility of the water molecules in aqueous sucrose solutions studied by ^{2}H NMR relaxation, *Z. Naturforsch.*, 1994, **49c**, 250–257.
157. R.H.M. Hatley, F. Franks, S. Brown, G. Sandhu and M. Gray, Stabilization of a pharmaceutical drug substance by freeze-drying: a case study, *Drug Stability*, 1996, **1**, 73–85.
158. R.H.M. Hatley, F. Franks and M. Green, A novel data acquisition, retention and examination system (DARES) for differential scanning calorimetry, *Thermochim. Acta*, 1989, **156**, 247–257.
159. N.A. Williams and T. Dean, Vial breakage by frozen mannitol solutions: correlation with thermal characteristics and effect of stereoisomerism, additives and vial configuration, *J. Parenter. Sci. Technol.*, 1991, **45** 94–100.
160. F. Franks, R. Ramanujam, J. Heaster, C. Huang, J. Jolly, J. Koelbl, C. Lively, E. Ogutu, E. Ting, S. Trembl, B. Aldous, R. Hatley, S. Mathias and B. Burdick, Ambient temperature stable molecular biology reagents, *Biotechniques*, 1993, **14**, 470–474.
161. F. Franks, R.H.M. Hatley and S.F. Mathias, 1995. Process for storage of materials. US Patent 5,928,469.
162. F. Franks, Long-term stabilization of biologicals, *Bio/Technology*, 1994, **12**, 253–256.
163. C.J. Roberts and F. Franks, Crystalline and amorphous phases in the binary system water–β,β-trehalose, *J. Chem. Soc. Faraday Trans.*, 1996, **92**, 1337–1343.

164. K.J. Crowley and L. Martini, Drug-excipient interactions, *Pharm. Technol. Europe*, 2001, **13**, 26–34.
165. J. Dubochet, M. Adrian, J.J. Chang, J. Lepault, A.W. McDowall and P. Schultz, Cryoelectron microscopy of vitrified specimens, *Quart. Rev. Biophys.*, 1988, **21**, 129–228.
166. F. Franks, T. Wakabayashi and S.F. Mathias, Nucleation kinetics in undercooled yeast cells: long-term stability against freezing, *J. Gen. Microbiol.*, 1987, **133**, 2807–2815.
167. M. Karel, M.P. Buera and Y. Roos, Effects of glass transitions on processing and storage, In *The Glassy State in Foods*, J.M.V. Blanshard and P.J. Lillford (eds.), Nottingham University Press, 2003, pp. 13–34.
168. C. Körber, M.W. Scheiwe and K. Wollhöver, Solute polarization during planar freezing of aqueous salt solutions, *Int. J. Heat Mass Trans.*, 1983, **26**, 1241–1253.
169. E. Mayer, New method for vitrifying water and other liquids by rapid cooling of their aerosols, *J. Appl. Phys.*, 1985, **58**, 663–667.
170. M.J. Pikal, Use of laboratory data in freeze-drying process design: heat and mass transfer coefficients and computer simulation of freeze drying, *J. Parenter. Sci. Technol.*, 1985, **39**(3), 115–139.
171. J. Pouchly, J. Biroš and S. Beneš, Heat capacities of water swollen hydrophilic polymers above and below 0°C, *Makromol. Chem.*, 1979, **180**, 745–760.
172. E.Yu. Shalaev and F. Franks, Equilibrium phase diagram of the water–sucrose–NaCl system, *Thermochim. Acta*, 1995, **255**, 49–61.
173. D. Turnbull, Under what conditions can a glass be formed?, *Contemp. Phys.*, 1969, **10**, 473.
174. G.M. Wang and A.D.J. Haymet, Trehalose and other sugar solutions at low temperature; modulated differential scanning calorimetry (MDSC), *J. Phys. Chem. B*, 1998, **102**, 5341–5347.
175. G.R. Wood and A.G. Walton, Homogeneous nucleation kinetics of ice from water, *J. Appl. Phys.*, 1970, **41**, 3027–3036.
176. M. Yoshioka, Y. Aso, K. Izutsu and T. Terao, Application of accelerated testing to shelf-life prediction of commercial protein preparations, *J. Pharm. Sci.*, 1994, **83**, 454–456.

Subject Index

activation barrier 7–9, 55–6, 66, 99–101
activity, international unit of (IU) 4
alditol 70
alternatives to freeze-drying
  currently used methods, 182–3
  evaporative drying, 183–6
amorphisation
  see vitrification
amorphous solids
  see glasses
aqueous PHC solutions
  complex dynamics 149, 161
  decoupled rotation 161–4
Arrhenius kinetics 30, 42, 63–6, 83
  at sub-T_g temperatures 162, 168
  deviations from 33
  relations to VTF equation 148–9

bioinactivation
  see stability

case studies
  formulation choice 173
  inadvertent failure 175–8
  process cycle optimisation 173
  product and process refinement 173–5
  product development process 171–2
  storage, *in situ* water release 179–80
  unfamiliarity with freeze-drying parameters 178–9
chill
  deep chill (*see* undercooled water (chilled))
  vs freezing 29–30
"clean-in-place" (CIP) system 14
Clostridium botulinum 12
collapse 97, 122, 127, 158, 168, 175–6, 179
computer simulations
  of freezing 24
  of pharmaceutical glasses 76
concentration damage 58–60
configurational entropy (S_c) 147, 150
containers
  blowback 134–5
  closure and packaging 134–5
  contamination 139–40
  dimensions 23, 179
  filling 130–4
  shape 106, 108–9, 116, 133–4
  solution fill depth 16, 109–11, 114–17, 124–5, 152, 172
  stoppers and overseals 134–5, 136
  volume 23, 172, 179–80
cooling rate 61–3, 84–7, 113–14, 124–5, 151
crystallisation
  amino acids 71
  Avrami equation 45
  from supersaturated solution 43–8
  hydrates 48–50, 145
  kinetics 36–9
  nomenclature 70
  of solutes from aqueous solution 43–8, 61–3
  peritectic transitions 43–4, 50, 51, 60–1
  perturbed 42
  primary 67
  secondary 67
  time-temperature-transformation (TTT) curve 45–6
  see also ice nucleation

de Veer, Gerrit 2–3
differential scanning calorimetry (DSC) 121

artefacts 154
deconvolution 166
estimation of shelf life 159–61
estimation of thermophysical parameters 84–7, 159–61
experimental details 151–4
interrupted 154
isothermal mode 160–1
mannitol-NaCl 70–1
modulated 154
of stored products 145–6, 152–4
residual water estimation 165–6
scanning rate 152, 155–7
water-mannitol-NaCl 70–1
water-sucrose-NaCl 68–70
drip loss 3
drying
as a preservation method, history 1–3
definition 17
overdrying 17–18
primary (*see* ice sublimation)
secondary (*see* secondary drying)

electron microscopy
batch failure 176
phase separation 29, 49, 69
enzyme activity, international unit of (IU) 4
equilibrium constant *(K)* 9
erythrocytes, human 53
eutectic point (T_e) 43–4, 57
evaporative drying
current practice 184–5
vs freeze-drying 185–6
principles and practice 183–4
excipients
approved 90
attributes 90–5
chemical damage 100–1
classification 89–90, 91
definitions 89
devitrification 99–100
effect on pH 91–2
functions 91
glassy 100–1
lyostabilising 92–3, 101
operational degrees of freedom 103–4
patent protection 90
peptides 93, 101–2
PHCs as favourites 90, 93–5
polymers 90, 102
product interaction 96–7
regulatory requirements 101
release of water 104
in vivo lyostabilisers 104
volatile 92

Ficoll 102
fill depth, containers 16, 109–11, 114–17, 124–5, 152, 172
Food Polymer Science concept 6
freeze concentration 2, 28–9, 58
and reaction rate enhancement 64–7, 87
freeze drier
capacity 4, 15, 130–4, 172, 179
chamber pressure 25
components 13–16
condenser 24–5
cross contamination 15
recorder output 25
validation 11
freeze-drying
accelerated 117–19
advantages and disadvantages 10–11, 18–19
automation 130
books 4
complex biological materials 52–3
deleterious processes 3
economics of 23, 131–4
energy requirement 18–19
excipients 97–9
formulation development 21–4
history 1–6
multidisciplinary nature of 24–6
organic solvents 118–19
of foods 1, 5, 9
packaging 134–5
scale-up 130–1, 132
vial closure 134–5
see also ice sublimation (primary drying)
vs evaporative drying 185–6
freeze-drying cycle 13–19
degrees of freedom 103–4
duration 16
industrial process 14

optimisation 17–18, 104
process stages 15
freezing 28–30
vs chill 29–30
freezing damage (injury) 58–60

glass fragility, experimental determination of 148–50, 154–8
glass transition (T_g)
enthalpy relaxation 82–5
estimation from melting point 84
of composites 79–81
origin 74
properties of excipients 94–6
relationships with molecular structure 83
semi-empirical estimation 95–6
sucrose-water 73
temperature (T_g) 72, 87–8, 105–6, 144, 151
width 157–8
glasses
as undercooled liquids 74–5
characterisation of 74–7, 78–82
chemical reactions in 54, 64–7
coexistence with crystals 79
comparison with crystals 78–82, 98
dynamics in 82–4
fragility parameter 148–50, 154–8
freezing behaviour 87–8, 97–9
inherent instability 77
methods of formation 71–7, 93
non-equilibrium processes in 82–4
quasi-equilibrium 82
safe storage temperature 97
structural relaxation rates 84–5
thermophysical properties 84–7
Gordon-Taylor equation 95–6, 167
Greaves, R.I.N. 5

haemolysis 52–3
salt and freezing equivalence 53, 59
heat transfer 107–12
by conduction 107–8
by convection 107–9
by radiation 107–8
coefficient K_v 16, 108, 109–12, 172
coupled with mass transfer 105, 106
human blood plasma 5–6, 59

ice
amorphous (*see* glasses)
crystal growth rate 113
crystallisation and crystal habits 42–3
equilibrium melting point (T_m) 34, 40
maturation (annealing) 42, 47–8, 113
perturbed crystallisation 42
polymorphism 48–50
sublimation pressure 34
ice nucleation
ecological implications 41
equilibrium freezing point (T_m) 34, 40
heterogeneous (T_{het}) 41
homogeneous (T_{hom}) 35, 36, 37–41
measurement of 37
nomenclature 41
rate *(J)* 36–7
relation to melting 34, 39
solute behaviour 43–4
substrate efficiency 41
theories of 33–42
see also crystallisation
ice sublimation (primary drying)
accelerated by tert-butanol (TBA) 117–19
cake porosity 109, 116
concentration gradients 113, 117
concentration polarisation 113
crystal size 113
driving force 107
effect of chamber pressure 106–7, 109, 114–16
effect of surface area 109
ideal conditions 117
interplay of processing parameters 114–16
latent heat of sublimation 106, 108, 114
modelling 105–7
monitoring 119–20
optimum conditions for 105–6
rate calculation 108, 109
resistance to 109–10, 113–14
saturation vapour pressure 107, 108, 111–12
solid content 113, 116
see also freeze drying
inadvertent failure
case studies 175–8

causes 146, 176, 177
investigation into 176–8
instability
see stability
instantisation technology 1
International Conference on Harmonisation of Technical Requirements (ICT) 131
isothermal calorimetry
limitations 144
in long-term stability measurement 144
sensitivity 154, 159–60
isothermal calorimetry, advantages over DSC 151, 167
isoviscosity profiles 148–9

lactic acid bacteria 5
Lobelia telekii 41
lyoprotectant(s) 23, 92–3
method of action 23, 92–3, 100–1, 104

mannitol 61, 70–1, 144
mannitol-NaCl (DSC) 70–1
water-mannitol-NaCl (DSC) 70–1
mass transfer 16, 105
and solid content 116–17
condenser capacity 107–8
coupled with heat transfer 105, 106
effect of cake surface area 107–9
effect of chamber pressure 105–16
effect of vial fill depth 109–11, 116–17
monitoring methods and devices
anemometer 120
mass spectrometry 120, 139
pressure gauge 120
pressure rise test 119–20, 137
process monitoring and control 135–8
rheology 151
temperature probe 119, 137
thermal analysis (DTA, DSC) 145–6, 152–4
thermogravimetric analysis (TGA) 151
thermomechanical analysis (TMA) 151
viscometry 147–9

neutron scattering, glucose 76, 146
Nicotiana (tobacco) embryo germination 53
nuclear magnetic relaxation, sugar-water mixtures 164
nutroceuticals 5

overdrying 17–18

pH, protein inactivation 8–9
pH buffers
buffering ranges 92
cooling effects 32–3, 63–4
crystallisation 60–4
excipients effects 91–2
phase separation 60
pharmaceutical glasses, as solid solutions 74–7
pharmacokinetics 26
phase equilibria
complex 60–4, 67–71
eutectic crystallisation 60–4
eutectic data 57–63, 67–75
eutectic point (T_e) 43–4, 57
multicomponent 67–71, 74–7
phase separation (polyamorphism) 77
rates of achievement (relaxation) 147–9
simple 28–30
solid-liquid 28–9, 43–8, 51, 57–60, 142
transition, 37–9
phosphate-buffered saline solution (PBS) 24, 61
Pikal, Michael J. 5
polyhydroxy compounds (PHCs)
as solid solvents 93–5, 99–101, 168–9
crystal structures 70–1
crystallisation 44, 63, 72
dynamics, concentration dependence 163–4
glass transitions 75–7
hydrate formation *in situ* 79, 145
hydrates 79, 145
hydrogen bonded networks 72–6
mannitol polymorphism 70–1
polymorphic transitions 48–50, 145
primary drying
see ice sublimation
probiotics 5

process
  cycle optimisation, case studies 173
  flow sheet 20–7
  parameters 20–7
product, dried
  development process 171–2
  physical state 141–4
  process refinement 173–5
  reconstitution 170
  shelf life determinants 145–50, 159–61
  stability 144–6
  ternary state diagram 142–3
  yield 21
protein
  aggregation and precipitation 61, 63
  cold inactivation 8–9, 55–7, 66–7
  denaturation 8–9, 55–6
  purification 21–2
  purity 21
  rDNA (r-proteins) 22
  solubility 56
  stability thermodynamics 55–7

raffinose pentahydrate
  amorphisation 79, 81, 146
  dehydration 121, 145–6
rDNA proteins (r-proteins) 22
relaxation
  Adam-Gibbs equation 150
  below T_g 146
  configurational entropy (S_c) 147, 150, 155
  configurational heat capacity 150
  Kauzmann temperature (T_k) 147
  molecular mobility 147–9
  softening point (*T*s) 154
  structural 146
  supersaturated solutions 161–4
  temperature of zero mobility (T_o) 147
  unresolved issues 168–70
  viscosity 147–9
  Vogel-Tammann-Fulcher (VTF) equation 147–9
residual water 13, 18, 164–7
  chemical reactivity 165, 168–70
  deleterious effects of 18, 165
  effect on stability 168
  estimation 164–7
    DSC scans 165–6
    Karl Fischer techniques 166–7
  as plasticiser 165
  promotion of crystallisation by 145, 161–9
  regulatory aspects 17–18
rheology 151
  of pharmaceutical glasses 151

Saccharomyces cerevisiae 53
secondary drying
  aggregation during 123, 126–7
  desorption 121, 123
  diffusion 123
  fill depth 124–5
  hydrate decomposition 121–2
  operational protocols 128–9
  overheating and collapse 125
  pressure effects 123
  process control 128–9
  softening point (*T*s) 127, 154
  surface area 124–5
  temperature
    effects 125–7
    ramping 128–9
  unfrozen water 121–3
  water content effects 123–4
solids, crystalline *vs* amorphous 78–82, 98, 141–50
stability 6–10, 139–40, 158
  accelerated assays 159
  browning reactions 7
  chemical 8–10, 158
  definitions 6–7
  dried product 144–6
  effect of residual water 168
  estimation of 151, 154, 159–60
  regulatory requirements (USA and Europe) 139
  self-stabilisation 146
  testing (ICH) 131
  thermomechanical 146–8, 151, 168
state diagram(s)
  definitions 72

importance 71–3
isoviscosity profiles 148–9
sterile operations 10, 40, 131–2
"sterilise-in-place" system 14
sublimation 4
sucrose
sugar-water mixtures, nuclear magnetic relaxation 164
water-sucrose-NaCl (DSC) 68–70
supersaturated solution(s)
and protein inactivation 55–7
crystallisation 43–8
dynamics 161–4

temperature
fictive $T_{f(T,t)}$ 84, 150
freezing T_f 13
glass transition (T_g) 72, 87–8, 105–6, 144, 151
Kauzmann temperature (T_k) 147
low (chill) 28–31
measurement of 119, 137
monitoring methods and devices 119, 137
product 20, 24–5, 106, 109–19, 126, 135–8, 168, 172
ramping, secondary drying 128–9
shelf 24, 115–19, 129, 138, 172–3, 251–09
temperature–composition state diagram 142–3
zero mobility (T_o) 147
thermoanalytical techniques 151–4

undercooled water (chilled)
as storage medium 31, 53
kinetics in 31
physical properties of 32–3
physicochemical properties 54–5
state diagram importance 71–3
uses of 53
see also ice nucleation
undercooling 28–30

Veer, Gerrit de 2–3
vitrification 44–5, 50–2
concentration 64–7, 87
desiccation 100
quench cooling 80, 82, 87, 94

water
amorphous ice (*see* glasses)
"bound" 5
dynamics, affected by sugars 161–4
equilibrium freezing point (T_m) 34, 40
ionisation equilibrium (pK_w) 32, 54
plasticizing effect of 165–7
release during storage 179–80
residual 121–3
water vapour pressure–temperature relationship 138
see also residual water; undercooled water
white chuño 2–3
Williams-Landel-Ferry (WLF) equation, stability calculations 158
Wollaston, William Hyde 4

X-ray diffraction
mannitol hydrate 61, 103
sucrose 68, 70

CPSIA information can be obtained
at www.ICGtesting.com
Printed in the USA
2751LVUK00001B

9780854042685